THE LEGACY OF IRON AND INNOVATION

The Legacy Of Iron And Innovation

Alina Hazel

UNIEK ENTERPRISES

Contents

INDEX

Introduction

In the records of mankind's set of experiences, scarcely any materials have made as permanent an imprint as iron. Its importance rises above ages and societies, creating a long and persevering through shaded area on the course of development. Iron, with its astounding properties, has woven itself into the texture of our lives, changing businesses, profoundly shaping social orders, and adjusting the direction of history. The tradition of iron and its creative applications is a significant demonstration of human resourcefulness and versatility.

Iron, perhaps of the most bountiful component on The planet, has for quite some time been an indispensable piece of human life. Its metals, tracked down in nature, have been refined and fashioned into devices, weapons, and decorations for centuries. The Iron Age, a crucial part in our story, denotes the change from the previous Bronze Age and proclaimed another period of material culture, financial frameworks, and mechanical advancement. This change in materials reshaped the world as well as manufactured a heritage that keeps on impacting our lives in the 21st 100 years.

At the core of this inheritance lies the extraordinary force of iron. This metal, with its outstanding strength, pliability, and sturdiness, plays had a focal impact in shaping social orders and driving advancement. From the development of notable designs like the Eiffel Pinnacle and the Brooklyn Scaffold to the improvement of progressive transportation frameworks and the introduction of the advanced modern age, iron's flexibility has been outfit in endless ways. It has filled our minds and permitted us to dream past the restrictions of the conceivable.

The tradition of iron reaches out a long ways past its utilitarian purposes. In the domain of workmanship and configuration, iron has been a material for imagination and articulation. Ironwork, from complicated created iron entryways to fabulous figures, has decorated our urban communities and organizations, adding excellence and character to our constructed climate. The combination of workmanship and iron stands as a demonstration of the human motivation to change the commonplace into the exceptional, a tradition of magnificence and craftsmanship that perseveres.

The excursion of iron advancement has not been restricted to the actual world alone. The synthetic and metallurgical cycles that have opened its true capacity have driven logical revelation and mechanical headway. From the leap forwards of the Modern Upheaval to contemporary advancements in materials science, the quest for understanding and bridling iron's properties has driven development across disciplines.

Iron's inheritance, nonetheless, isn't absent any trace of intricacies. The extraction and handling of iron metals have presented ecological difficulties, from deforestation to contamination. The interest for iron has prompted asset consumption and international strains. Besides, the effect of iron creation on work conditions and social designs has been a subject of progressing concern.

The tradition of iron is a nuanced embroidery, woven with the two victories and hardships, advising us that progress frequently includes some major disadvantages.

As we dig into the tradition of iron and advancement, taking into account its multi-layered impact on our world is basic. This investigation will navigate the Iron Age and its verifiable importance, the specialized parts of iron creation and handling, the different utilizations of iron in industry and framework, the imaginative articulations through ironwork, and the ecological and cultural difficulties presented by its proceeded with use. Through this diverse focal point, we can acquire a more profound enthusiasm for the significant and getting through effect of this momentous component.

The Iron Age: A Urgent Age

To fathom the tradition of iron and development, we should travel back to when iron initially arisen as a groundbreaking power — the Iron Age. Albeit the Iron Age unfurled at various times in different locales of the world, it remains as a vital age in mankind's set of experiences.

The Iron Age addresses a takeoff from the previous Bronze Age, portrayed by the inescapable utilization of bronze, a copper-tin composite. Bronze had its benefits, however iron had remarkable properties that would end up being a distinct advantage. Iron is plentiful in nature, which implied that it was promptly accessible and moderately reasonable. Be that as it may, its actual importance lay in its actual properties — iron was more earnestly, more solid, and more promptly honed than bronze. These characteristics made iron an optimal material for devices, weapons, and a great many executes.

Verifiable records and archeological proof propose that the Iron Age started in different regions of the planet at various times. In Anatolia, current Turkey, it is accepted to have started around 1200 BCE, while in Focal Europe, it arose somewhere in the range of 800 and 600 BCE. The Iron Age in India started around 1200 BCE, and in sub-Saharan Africa, the course of events fluctuates by locale, for certain region entering the Iron Age as soon as 2000 BCE. As ironworking strategies spread, they reshaped the material culture as well as the social, monetary, and political designs of the social orders that embraced them.

The shift from bronze to press was not only an innovative progress; it denoted a huge social change too. Iron apparatuses and weapons significantly improved agrarian works on, speeding up food creation. This excess of food prompted populace development and the advancement of additional mind boggling social orders. Iron's part in the creation of weaponry additionally had significant ramifications for fighting, influencing the overall influence and successes. Basically, the Iron Age denoted the ascent of realms and the reconfiguration of political scenes.

Maybe one of the most notable and getting through traditions of the Iron Age is the iron furrow. This development altered farming by making it more effective and useful.

The expanded yields took into consideration bigger populaces and the rise of metropolitan focuses, as well as the specialization of work. The iron furrow, with its capacity to work soil all the more actually, represented the extraordinary force of this new material.

The meaning of the Iron Age isn't bound to a solitary age; its impact continued and developed over the long haul. As social orders progressed and innovation advanced, iron kept on assuming a focal part in their turn of events. The tradition of the Iron Age, with its underlying foundations in agribusiness and fighting, established the groundwork for the ensuing developments and changes that would shape the course of human progress.

The Metallurgical Authority: Releasing Iron's True capacity

The tradition of iron's development reaches out a long ways past its verifiable rise in the Iron Age. The authority of metallurgy, or the study of working with metals, has been instrumental in opening iron's true capacity and driving mechanical advancement. Iron, in its crude structure, isn't promptly usable. It should go through a progression of intricate metallurgical cycles to be changed into the flexible material we know today.

The way in to press' change lies in its extraction from iron minerals, a cycle that has developed fundamentally throughout the long term. Iron minerals, like hematite and magnetite, are bountiful in the World's covering. The test lies in isolating the iron from the debasements, a cycle known as refining. Purifying includes warming the mineral to high temperatures within the sight of a decreasing specialist, typically charcoal or coke, to remove the iron. This cycle lets the iron out of the oxygen in the mineral, bringing about liquid iron.

By and large, early iron purifying tasks were limited scale issues, frequently utilizing bloomeries, which created "blossoms" of iron that were then manufactured into usable shapes. Over the long run, these procedures advanced, prompting the improvement of bigger and more effective heaters. The impact heater, a vital creation, changed iron creation during the late middle age time frame.

The impact heater, fueled by constrained air, took into consideration the high-temperature decrease of iron metal on a modern scale. This advancement essentially expanded the yield of iron and decreased creation costs. The boundless reception of shoot heaters during the Modern Upset denoted a defining moment throughout the entire existence of iron creation. This period saw the advancement of metallurgical procedures that would establish the groundwork for present day steelmaking.

Steel, a composite of iron and carbon, addresses one more section in the metallurgical dominance of iron. By controlling the carbon content and utilizing different intensity treating processes, steelmakers could make materials with many properties, from delicate and pliable to hard and solid.

Steel turned into the foundation of the Modern Upset, filling in as the establishment for extensions, railroads, and hardware. The tradition of steel and its continuous importance in the advanced world are demonstration of the force of metallurgy in molding our material culture.

The authority of metallurgy stretched out past the production of iron and steel. Advancements in alloying, or consolidating iron with different components, delivered materials with excellent properties. For example, the expansion of chromium to press made hardened steel, which is profoundly impervious to erosion. These metallurgical headways have had expansive effects, in development and designing as well as in the improvement of particular materials for aviation, medical care, and endless different enterprises.

The advancements in metallurgy and press handling have expanded the productivity of creation as well as extended the scope of uses for iron and its compounds. This extension has reshaped the world, considering the development of always aggressive designs and the advancement of perplexing apparatus that drives our cutting edge lives. The dominance of iron has been a main thrust behind the perpetual walk of human advancement.

1. **Setting the Stage**

   In the always advancing show of human progress, the stage is never-endingly set for the unfurling of stories that characterize our past, shape our present, and move us into what's to come. This stage envelops the huge territory of our planet, packed with topographical highlights, biological systems, and climatic varieties. It is the material whereupon the complicated embroidery of mankind's set of experiences, culture, and advancement is woven. To comprehend the complexities of our reality and the difficulties we face, investigating the assorted scenes and natural settings that set up for the human story is fundamental.

   Geology, in its broadest sense, is the groundwork of our reality. It incorporates the actual elements of the Earth, from transcending mountain reaches to immense seas, from rich rainforests to dry deserts. These topographical highlights, molded north of millions of years by geographical powers, have been major in characterizing the limits and conceivable outcomes of human social orders. Geology is, generally, the material whereupon our story unfurls, and the background against which our accomplishments and battles are compared.

Quite possibly of the most striking geological component on our planet is the tough territory of mountain ranges. Mountains play had an essential impact in molding mankind's set of experiences. From the strong Himalayas to the Andes, these geographical behemoths have gone about as the two hindrances and channels for human development. Mountain ranges have, now and again, separated networks, prompting particular societies and dialects, while additionally giving essential assets like minerals, water, and prolific land. The difficulties of living in rugged areas have encouraged flexibility and versatility, as networks have needed to fight with steep slants, cruel environments, and restricted admittance to assets.

On the other hand, marsh fields and waterway valleys have additionally been critical in the human account. The ripeness of stream valleys, like the Nile, Tigris and Euphrates, and the Indus, has empowered the improvement of mind boggling agrarian social orders. These districts turned into the supports of antiquated civilizations, where agribusiness and exchange flourished. The yearly flooding of these stream valleys kept supplement rich residue, giving a consistent wellspring of food for prospering populaces. The extraordinary rural transformations that noticeable key defining moments in mankind's set of experiences, similar to the Neolithic Unrest, were firmly connected to these prolific marshes.

Our reality is additionally accentuated by immense territories of desert. Deserts, like the Sahara, the Bedouin Desert, and the Gobi Desert, present one of a kind difficulties to human home. Their unforgiving and bone-dry circumstances request variation and development for endurance. Many desert societies have flourished by outfitting restricted water assets, creating proficient water system frameworks, and becoming amazing at desert agribusiness. In these apparently ungracious conditions, antiquated shipping lanes and train urban areas thrived, connecting different locales of the world.

As we move from dry deserts to rich rainforests, we experience one more feature of the World's geological variety. Rainforests, like the Amazon, the Congo, and Southeast Asian rainforests, are unrivaled repositories of biodiversity. Their natural extravagance upholds a variety of greenery, large numbers of which stay unseen. Native societies, profoundly interlaced with these conditions, have drawn food from rainforests for centuries. Notwithstanding, the sensitive harmony between human exercises and rainforest protection has turned into a squeezing worry as deforestation and territory obliteration undermine these indispensable biological systems.

Islands, dissipated across the seas, present one more geological setting. These disconnected expanses of land have brought about novel biological systems and societies. Island social orders, whether in the Pacific, the Caribbean, or the Mediterranean, have developed unmistakable practices, dialects, and ways of life. Islands, by their actual nature, cultivate a healthy identity dependence and relationship, as assets are restricted and admittance to the central area is frequently difficult. The connection between island conditions and human social orders has created rich stories of investigation, marine, and social variety.

Streams and water bodies have been fundamental to the human experience, filling in as the two life savers and boundaries. Waterways, like the Yangtze, the Mississippi, and the Danube, have worked with transportation, exchange, and correspondence. They have supported agrarian social orders, empowering the improvement of intricate settlements and exchange organizations. The job of waterways in mankind's set of experiences is maybe best exemplified by the Nile, which not just upheld the thriving civilization of antiquated Egypt yet additionally molded its strict convictions and social personality.

Seas, covering over 70% of the World's surface, have given roads to investigation, exchange, and success. Seas have associated mainlands through sea courses, taking into consideration the trading of merchandise, thoughts, and people groups. The Time of Investigation, set apart by the journeys of Christopher Columbus, Ferdinand Magellan, and Zheng He, extended the skylines of human information and catalyzed the Columbian Trade, which changed worldwide exchange and environments. Seas, notwithstanding, have additionally been hindrances, isolating landmasses and restricting contact between social orders.

The World's topographical variety has affected the scattering of societies and the advancement of dialects. Disengagement, whether forced by mountain ranges or immense seas, has prompted the advancement of novel lingos and dialects. The etymological embroidery of the world is an impression of these geological limits and the restricted exchange between disconnected networks. The enormous biodiversity found in different biological systems has likewise been a wellspring of information and motivation, as native people groups have fostered a profound comprehension of their surroundings and the plants and creatures that possess them.

The effect of geology isn't restricted to regular scenes alone. It reaches out to the political and monetary domains. Borders, frequently portrayed by geological elements, have characterized the regional degree of countries and realms. Arguments about asset rich domains, like stream bowls or mineral-rich mountains, have filled clashes since the beginning of time. Monetary exercises have additionally been significantly affected by geology, as admittance to assets, shipping lanes, and environment conditions have decided the financial fortunes of locales and countries.

The interaction among topography and environment is a basic part of our reality's stage. Environment designs, impacted by the World's pivotal slant and position comparative with the sun, significantly affect biological systems, farming, and human residence. The equator, for instance, encounters reliably warm temperatures, prompting tropical rainforests, while the polar locales are described by outrageous virus. Occasional varieties in temperature and precipitation designs have formed agrarian works on, influencing the sorts of yields that can be developed and reaped.

The unique idea of our planet's environment is proven by peculiarities like El Niño and La Niña, which impact weather conditions and horticultural results. These climatic occasions, driven by connections between the sea and the air, can prompt dry spells, floods, and outrageous climate occasions, affecting food security and vocations. As the world wrestles with environmental change and its ramifications, the meaning of understanding the complex connection among topography and environment turns out to be progressively apparent.

Cataclysmic events, one more feature of the World's stage, have left permanent engravings on human social orders. Seismic tremors, volcanic ejections, storms, and tidal waves have formed scenes and affected human settlement designs.

The obliteration fashioned by these occasions has constrained social orders to adjust and devise systems for moderating their effect. The flexibility and creativity of networks confronting catastrophic events stand as a demonstration of the human limit with respect to endurance and recuperation despite misfortune.

2. **The Significance of Iron in Human History**

Iron, quite possibly of the most plentiful component on The planet, has significantly impacted mankind's set of experiences, forming the course of civilization in momentous ways. Its importance in the chronicles of humankind can hardly be exaggerated, as iron has been both an impetus for development and a foundation of mechanical advancement. The tale of iron's job in mankind's set of experiences is a demonstration of our ability for resourcefulness, versatility, and change.

Iron's verifiable noticeable quality is established in the progress from the Bronze Age to the Iron Age. The Iron Age, set apart by the coming of iron-working procedures, addressed a vital change in material culture. Before this change, social orders basically depended on bronze, a copper-tin compound, for instruments, weapons, and executes. While bronze had its benefits, it was iron's uncommon properties that would on a very basic level change the direction of human turn of events.

The Iron Age unfurled across different regions of the planet at various times, however it shared a consistent idea of innovative progression. In Anatolia (advanced Turkey), the Iron Age started around 1200 BCE, and in Focal Europe, it arose somewhere in the range of 800 and 600 BCE. India's Iron Age started around 1200 BCE, and in sub-Saharan Africa, it fluctuated by locale, for certain region entering the Iron Age as soon as 2000 BCE. Ironworking methods, borne of trial and error and advancement, spread across these areas, lighting changes in material culture, economy, and innovation.

Iron's predominance over bronze was multi-layered. It is promptly accessible in nature, making it a more open and financially savvy asset. However, it was iron's actual properties that really separate it. Iron is more diligently, more sturdy, and more promptly honed than bronze. It tends to be formed into instruments, weapons, and a horde of carries out no sweat. Iron's flexibility and strength made it an optimal material for a great many applications, and its acquaintance was associated with a mechanical upheaval.

The reception of iron apparatuses had significant ramifications for agribusiness, fighting, and cultural turn of events. Iron executes, like furrows, scrapers, and sickles, tremendously worked on farming practices. The iron furrow, specifically, was a distinct advantage, as it expanded the proficiency of soil development, prompting higher harvest yields. This excess of food prodded populace development and the advancement of additional mind boggling social orders. The capacity to take care of bigger populaces was a basic driver of social intricacy, empowering the rise of metropolitan focuses and the specialization of work.

Iron's part in the development of weaponry was similarly groundbreaking. The hardness and strength of iron made it ideal for creating weapons, which held huge influence in the old world. Iron weapons, like blades, lances, and protection, upset fighting, affecting the overall influence and triumphs. Social orders with admittance to prevalent iron weapons frequently acquired a critical benefit over their enemies. The Iron Age denoted the ascent of realms and the reconfiguration of political scenes.

The tradition of the Iron Age is established in the primary commitments of iron to farming and fighting. The horticultural excess it empowered was essential for food as well as for exchange and financial turn of events. Urban areas and states prospered around districts with admittance to press, adding to the ascent of metropolitan focuses and the foundation of coordinated states.

Iron's effect was not restricted to the actual domain; it impacted the otherworldly and social components of social orders. Old metalworkers, frequently viewed as talented craftsmans, held a venerated place in many societies. The specialty of ironworking became interlaced with folklore, legends, and strict convictions. Iron's change from crude metal to helpful executes reflected the catalytic cycles of change in the otherworldly domain, further hardening its importance in human culture.

The dominance of metallurgy, or the study of working with metals, assumed a significant part in bridling iron's true capacity. Iron, in its crude structure, isn't promptly usable; it should go through a progression of mind boggling metallurgical cycles to be changed into the flexible material we know today.

One of the vital stages in this cycle is the extraction of iron from iron metals. Iron minerals, like hematite and magnetite, are bountiful in the World's covering. The test lies in isolating the iron from the pollutions in these minerals, a cycle known as purifying. Refining includes warming the mineral to high temperatures within the sight of a decreasing specialist, generally charcoal or coke, to separate the iron. This cycle sets the iron free from the oxygen in the mineral, bringing about liquid iron.

By and large, early iron refining tasks were limited scale undertakings, frequently utilizing bloomeries, which created "sprouts" of iron that were then manufactured into usable shapes. Over the long run, these methods advanced, prompting the improvement of bigger and more effective heaters. The impact heater, an original development, changed iron creation during the late middle age time frame.

The impact heater, controlled by constrained air, took into consideration the high-temperature decrease of iron metal on a modern scale. This development fundamentally expanded the yield of iron and decreased creation costs. The far reaching reception of shoot heaters during the Modern Upheaval denoted a defining moment throughout the entire existence of iron creation. This period saw the advancement of metallurgical methods that established the groundwork for present day steelmaking.

Steel, a composite of iron and carbon, addresses one more part in the dominance of metallurgy. By controlling the carbon content and utilizing different intensity treating processes, steelmakers could make materials with a large number of properties, from delicate and pliable to hard and sturdy. Steel turned into the foundation of the Modern Unrest, filling in as the establishment for scaffolds, rail routes, and apparatus. The tradition of steel and its continuous importance in the cutting edge world are demonstration of the force of metallurgy in molding our material culture.

The advancements in alloying, or consolidating iron with different components, have created materials with excellent properties. For instance, the expansion of chromium to press makes hardened steel, which is profoundly impervious to erosion. These metallurgical headways have had expansive effects, in development and designing as well as in the advancement of particular materials for aviation, medical services, and endless different businesses.

Iron's dominance has risen above the domains of industry and framework, reaching out into the universe of craftsmanship and inventiveness. Iron, with its ability for change and control, has filled in as a material for imaginative articulation and craftsmanship all through the ages.

Fashioned ironwork, described by its complex plans and fancy examples, is maybe the most commended type of imaginative articulation through iron. This specialty has a rich history, with establishes in the Iron Age and the Mediterranean locale. Throughout the long term, it has developed into an exceptionally particular work of art, with ace metalworkers making elaborate doors, railings, and ornamental components.

The excellence of fashioned iron lies in its capacity to be manufactured and formed by talented craftsmans. Smithies utilize different methods, like turning, looking over, and punching, to make unpredictable and fancy plans. From the rich scrollwork of European iron railings to the fragile filigree found in Center Eastern engineering, fashioned iron has embellished structures, nurseries, and public spaces, adding a dash of tastefulness and appeal to the fabricated climate.

The utilization of iron in mold is one more demonstration of its true capacity as an imaginative medium. Iron figures, whether conceptual or illustrative, have graced public spaces, displays, and historical centers. Specialists like Alexander Calder and David Smith have investigated the conceivable outcomes of iron as a sculptural material, making striking and persevering through works that push the limits of structure and articulation.

The tradition of iron in craftsmanship stretches out to different types of imaginative articulation also. Iron has been utilized in the production of enhancing and useful articles, from ceiling fixtures and candelabras to furniture and family things. The combination of structure and capability, craftsmanship and configuration, has prompted the creation of articles that join stylish allure with useful utility.

Iron's job in imaginative articulation isn't restricted to the actual domain. The utilization of iron as an illustration and image has pervaded writing, verse, and narrating. Its solidarity and toughness frequently act as strong similitudes for human flexibility and determination. The tradition of iron in the domain of imagination advises us that craftsmanship and excellence can be tracked down in the most improbable spots, even in the unrefined components of industry and framework.

The tradition of iron and its creative applications in industry, foundation, and imaginative articulation highlights its getting through importance. The world as far as we might be concerned would be unrecognizable without the commitments of iron to development, transportation, assembling, and innovation. The adaptability and strength of iron have permitted us to fabricate, associate, and make in manners that were once unbelievable.

While the effect of iron in mankind's set of experiences is unequivocally sure in many regards, it is a story set apart by intricacies and difficulties. The extraction and handling of iron minerals, while fundamental for human advancement, have presented huge ecological worries, and the interest for iron has prompted asset exhaustion and international strains. The effect of iron creation on work conditions and social designs has likewise been a subject of progressing concern.

3. The Role of Innovation

Advancement, the innovative flow of imagining, creating, and carrying out groundbreaking thoughts, innovations, or techniques, is the main thrust behind human advancement. It shapes our reality by encouraging progressions in science, industry, workmanship, and society. Over the entire course of time, development has impelled mankind forward, prodding extraordinary changes and molding the manner in which we live, work, and communicate. This exposition investigates the focal job of development in mankind's set of experiences, with a specific spotlight on its effect in different spaces.

The Idea of Development

Development is an expansive and diverse idea that envelops a large number of exercises and results. It isn't restricted to a particular field or area yet rather works as a powerful power that penetrates each part of human life. Developments can be unmistakable, like an earth shattering innovation or another item, or theoretical, like clever thoughts, methods of reasoning, or social practices.

The course of development normally includes the recognizable proof of an issue, challenge, or opportunity, trailed by the age of clever fixes or thoughts to address it. These thoughts are then refined, created, and executed to deliver a significant change or progression. Advancement frequently requires cooperation and cross-disciplinary reasoning, as it draws on different subject matters and ability to track down original arrangements.

Advancement and Innovation

One of the most prominent spaces where development has had a significant effect is innovation. Since the beginning of time, people have persistently endeavored to foster new innovations that work on their personal satisfaction, increment efficiency, and extend their capacities. Mechanical development has been a main thrust behind logical revelations, financial development, and cultural change.

The historical backdrop of mechanical development is packed with game-changing innovations that have reshaped society. The print machine, developed by Johannes Gutenberg in the fifteenth 100 years, upset the scattering of information by making books more open to a more extensive crowd. The steam motor, created during the Modern Insurgency, fueled the hardware that impelled the large scale manufacturing of merchandise, changing economies and metropolitan scenes.

In the twentieth 100 years, advancements in processing and media communications upset how individuals convey and share data. The appearance of the web and the advancement of PCs introduced the computerized age, empowering the quick trade of information and the making of virtual universes. Developments in man-made brainpower and AI have additionally extended the outskirts of innovation, with applications going from independent vehicles to clinical diagnostics.

Mechanical advancement has changed ventures as well as affected cultural standards and ways of behaving. The improvement of cell phones, for example, has reshaped the manner in which individuals connect, work, and consume data.

These gadgets have become vital to present day life, empowering moment correspondence, admittance to an abundance of data, and a horde of utilizations and administrations.

Development in innovation has additionally been instrumental in tending to a portion of the world's most squeezing difficulties. Leap forwards in sustainable power advancements, for instance, have offered answers for moderate the impacts of environmental change by decreasing fossil fuel byproducts. Propels in clinical science have prompted the advancement of life-saving medicines and immunizations. Advancement in transportation, like electric vehicles and rapid trains, can possibly make transportation more manageable and effective.

The Force of Logical Disclosure

Logical development, the most common way of procuring new information and grasping the normal world, has been a foundation of human advancement. Science is driven by interest and the longing to uncover the central standards administering the universe. Logical disclosure has prompted a more profound comprehension of the physical and organic cycles that support life on The planet.

In the domain of physical science, the revelations of figures like Sir Isaac Newton and Albert Einstein have altered how we might interpret the regulations overseeing the movement of articles and the basic standards of the universe.

Newton's laws of movement and general attraction established the groundwork for traditional mechanics, while Einstein's hypothesis of relativity changed how we might interpret space, time, and gravity.

In the area of science, advancements in the investigation of hereditary qualities and development have given significant experiences into the variety of life on The planet. Charles Darwin's hypothesis of development by regular choice reshaped how we might interpret how species adjust and advance over the long haul. The disclosure of the design of DNA by James Watson and Francis Cramp made the way for another time of hereditary exploration and biotechnology.

Developments in science extraordinarily affect society. The improvement of the intermittent table by Dmitri Mendeleev gave a structure to figuring out the properties of synthetic components and foreseeing the presence of unseen components. The creation of the polymerase chain response (PCR) by Kary Mullis reformed subatomic science and hereditary qualities by empowering the enhancement of DNA groupings.

Logical development isn't restricted to the research facility; it has broad applications in different spaces. Clinical examination, for instance, has prompted the improvement of life-saving immunizations, the disclosure of anti-toxins, and advances in careful methods. Farming science has further developed crop yields and added to food security. Ecological science has given bits of knowledge into the effect of human exercises in the world and directed endeavors to address environmental change and safeguard biodiversity.

Development in science is much of the time driven by interdisciplinary cooperation and the sharing of information and thoughts. The logical technique, a precise way to deal with request and trial and error, plays had a focal impact in propelling comprehension we might interpret the world. The distribution and dispersal of exploration discoveries have permitted researchers to expand on one another's work, speeding up the speed of revelation.

Creative and Social Advancement

Development isn't restricted to the domains of science and innovation; it likewise assumes a vital part in human expression and culture. Imaginative advancement includes the formation of new imaginative structures, styles, and articulations that challenge regular standards and grow the limits of human innovativeness.

Since forever ago, craftsmen, essayists, artists, and entertainers have pushed the limits of their individual mediums, presenting novel ideas and thoughts that incite thought and feeling. The Renaissance, a time of creative and social resurrection in Europe during the fourteenth to seventeenth hundreds of years, is a great representation of imaginative development. Specialists like Leonardo da Vinci, Michelangelo, and Raphael reformed painting and model, while journalists like Dante Alighieri and Geoffrey Chaucer made critical commitments to writing.

The twentieth century saw a flood in imaginative development, with developments like Cubism, Oddity, and Dynamic Expressionism testing customary creative shows. Advancements in photography, film, and computerized media have changed visual narrating, empowering new methods of imaginative articulation. The appearance of the web and virtual entertainment has democratized the dissemination of craftsmanship and permitted specialists to contact worldwide crowds.

In the domain of music, development plays had a focal impact in the advancement of melodic classifications and styles. Developments in instruments, recording advances, and sound designing have extended the potential outcomes of sound and creation. Figures like Ludwig van Beethoven, Jimi Hendrix, and John Coltrane are praised for their earth shattering commitments to music.

Social development reaches out to language, theory, and accepted practices. Rationalists like Immanuel Kant, Karl Marx, and Friedrich Nietzsche presented clever thoughts that tested winning convictions and philosophies. Essayists like Franz Kafka and George Orwell investigated tragic subjects that keep on reverberating with perusers. Social trend-setters, including social equality activists, women's activists, and backers for LGBTQ+ freedoms, have driven social change by testing laid out standards and pushing for fairness and equity.

Creative and social development is in many cases an impression of the social and political environment of the times. It can act as an impetus for social change, bringing issues to light of basic issues and motivating aggregate activity. Artistic expressions have the ability to rise above boundaries and societies, cultivating associations and understanding among individuals from different foundations.

Development In the public eye and Administration

Development likewise assumes a basic part in the improvement of cultural designs and administration. Social and political development envelops changes parents in law, foundations, and arrangements that mean to address arising difficulties and advance the prosperity of networks.

The improvement of vote based administration frameworks, for example, addresses a critical development throughout the entire existence of political idea. A majority rules system, with its standards of well known sway and delegate government, has been a main impetus behind friendly advancement and the security of individual privileges. The Magna Carta, endorsed in 1215, is viewed as one of the basic reports of vote based administration, as it laid out the rule that the lord is dependent upon law and order.

Advancements in administration have additionally reached out to the improvement of worldwide associations and arrangements. The Unified Countries, established in 1945, denoted an achievement chasing worldwide collaboration and the advancement of harmony and security. Peaceful accords like the All inclusive Announcement of Basic freedoms and the Paris Settlement on environmental change address cooperative endeavors to address worldwide challenges.

Chapter 1

Ancient Beginnings

Mankind's set of experiences is an embroidery woven over centuries, with each string addressing a novel story, culture, and period. Our mission to comprehend these old starting points is an excursion of disclosure that traverses the globe and arrives at back so as to the actual underlying foundations of human progress.

The tale of mankind's set of experiences starts in Africa, where our species, Homo sapiens, arose close to a long time back. It was in this immense and different mainland that our progenitors previously strolled upstanding, leveled up their apparatus making abilities, and fostered the mental capacities that would separate them from different primates. Over incalculable ages, they extended their reach, adjusted to various conditions, and gradually developed into the cutting edge people we are today.

As people spread out of Africa, they wandered into a world immensely not the same as the one we know today. During the last Ice Age, which arrived at its top about a long time back, a large part of the planet was shrouded in ice, and ocean levels were essentially lower. This permitted people to move to new grounds, including the Americas, Australia, and islands in the Pacific. These early trailblazers would confront difficulties and adjust to a great many conditions, from the cold tundra of North America to the deserts of Australia.

One of the most exceptional parts of antiquated starting points is the sheer variety of human societies that grew freely all over the planet. From the extraordinary human advancements of the Nile and the Tigris-Euphrates to the itinerant clans of the Eurasian steppes, every general public had its own interesting language, customs, and convictions. These societies thrived and here and there conflicted, abandoning a rich embroidery of customs and developments that keep on impacting our present reality.

In Mesopotamia, the support of human progress, the Sumerians made perhaps the earliest process for composing, known a cuneiform. This wedge-formed script, carved onto earth tablets, permitted them to record everything from strict texts to deals. It was an exceptional jump forward in human correspondence and established the groundwork for future composing frameworks.

In the mean time, in Egypt, the development of the pyramids exhibited the mind boggling designing ability of this antiquated progress. The Giza Pyramids, specifically, stay a demonstration of human resourcefulness, as researchers and archeologists keep on unwinding the secrets of their development and reason.

Toward the east, in the Indus Valley, the Harappan human advancement was prospering. Their high level metropolitan preparation, with efficient roads and seepage frameworks, indicates a refined society with an economy in light of exchange and farming. The Harappan script, nonetheless, stays undeciphered, leaving an enticing secret for history specialists.

The old civic establishments of the Americas were no less amazing. The Maya, for instance, fostered a complicated arrangement of hieroglyphic composition and an exact schedule. Their accomplishments in science, cosmology, and engineering are a demonstration of their scholarly and mechanical headways.

In old China, the Zhou Administration presented the idea of the Order of Paradise, which attested that the ruler's position was supernaturally appointed. This thought significantly affected the political way of thinking of East Asia into the indefinite future.

The ascent of the Roman Republic and later the Roman Realm denoted a critical crossroads in old history. The Romans extended their domain across Europe, North Africa, and the Center East, leaving a tradition of regulation, designing, and administration that lastingly affects Western progress. The development of streets, water passages, and stupendous design, for example, the Colosseum and the Pantheon mirrored their designing ability.

As we adventure further into the records of time, obviously the account of antiquated starting points isn't restricted to significant civilizations alone. More modest, less popular social orders likewise assumed huge parts in forming the course of history. The Olmec nation of Mesoamerica, for example, are frequently viewed as the "mother culture" of the Americas, with their particular epic stone heads and high level information on farming.

In West Africa, the realm of Ghana arose as a strong exchanging state, working with the trading of gold, salt, and different merchandise across the Sahara Desert. Ghana's abundance and impact prepared for future West African realms like Mali and Songhai.

The Silk Street, a huge organization of shipping lanes interfacing East and West, was one more essential improvement in old history. It encouraged the trading of merchandise, thoughts, and societies across Eurasia and then some. The transmission of innovations like papermaking, printing, and explosive had expansive results, reshaping social orders on the two finishes of the shipping lanes.

While these antiquated civic establishments and exchange networks flourished, the world's native societies, frequently disconnected from these worldwide turns of events, kept on advancing in their own special ways. The native people groups of the Americas, the Natives of Australia, and different African clans kept up with their rich practices and profound associations with the regular world.

The spread of religions likewise assumed a urgent part in deeply shaping old social orders. In the Indian subcontinent, the lessons of Siddhartha Gautama, known as the Buddha, led to Buddhism, a way of thinking that underlined the easing of enduring profound practice and moral living. In the Center East, the monotheistic confidence of Zoroastrianism arose, with its attention on the timeless battle among great and wickedness.

Judaism, one of the world's most seasoned monotheistic religions, had its foundations in the antiquated Israelite individuals. Their strict texts, including the Torah, keep on affecting the convictions and practices of millions of individuals today.

The rise of Christianity in the Roman territory of Judea denoted a significant crossroads in strict history. The life and lessons of Jesus Christ, as kept in the New Confirmation, pulled in a following that in the end prompted the spread of Christianity all through the Roman Domain and then some. The early Christian church confronted abuse however persisted, in the long run turning into the authority religion of the Roman Realm in the fourth hundred years.

In the seventh hundred years, the Bedouin Landmass saw the ascent of Islam, established by the Prophet Muhammad. The Quran, the blessed book of Islam, turned into a wellspring of direction and motivation for a huge number of individuals. The Islamic Realm quickly extended, bringing novel thoughts, sciences, and a rich social legacy to tremendous districts of Asia, Africa, and Europe.

All through these strict turns of events, clashes and epic showdowns frequently emerged. The Campaigns, a progression of strict conflicts among Christians and Muslims in the Center East, were a powerful illustration of how confidence and governmental issues entwined in the old world. These conflicts left an enduring effect on the districts in question, molding the connections among societies and religions for quite a long time into the future.

The antiquated world was not exclusively characterized by wars and clashes, nonetheless. It was likewise a period of incredible scholarly and imaginative thriving. In antiquated Greece, rationalists like Socrates, Plato, and Aristotle established the underpinnings of Western way of thinking. Their investigations into the idea of the real world, morals, and governmental issues keep on affecting philosophical idea right up to the present day.

Greek craftsmanship and engineering, with their accentuation on equilibrium and extent, made a significant imprint on resulting imaginative developments. The

Parthenon, a sanctuary devoted to the goddess Athena, remains as a demonstration of the Greeks' design accomplishments.

The Greek time frame that followed the triumphs of Alexander the Incomparable saw the combination of Greek and Eastern societies. The Library of Alexandria, established in Egypt, turned into a focal point of learning and grant, drawing in probably the best personalities of the time. It was here that information from across the explored parts of the planet was gathered, protected, and developed.

In antiquated Rome, designing wonders like water passages, streets, and extensions existed together with the glory of the Roman Discussion and the magnificence of mosaics and frescoes. Roman writing, with writers like Virgil and Ovid, significantly affects Western writing and folklore.

Further east, in India, the Maurya and Gupta Realms made huge commitments to arithmetic, science, and writing. The idea of nothing, the decimal framework, and central texts like the Arthashastra and Kamasutra began during this time.

Antiquated China saw momentous advances in innovation, especially in the fields of metallurgy, papermaking, and printing. The Incomparable Mass of China, a fantastic accomplishment of designing, was developed to safeguard against northern intrusions. Savants like Confucius and Laozi left getting through insight that impacted Chinese idea and culture for centuries.

In Mesoamerica, the Maya fostered a refined arrangement of hieroglyphic composition and a complicated schedule. Their insight into cosmology permitted them to make exact and perplexing schedules. The remnants of urban communities like Tikal and Palenque authenticate their structural accomplishments.

The fall of the Western Roman Domain in the fifth century denoted the start of the Dim Ages in Europe, a period described by flimsiness and the decay of brought together power. It was during this time that different Germanic and Viking clans cleared across the mainland, introducing another period of archaic history.

One of the most getting through traditions of the Dim Ages was the ascent of feudalism, a framework where land was conceded in return for steadfastness and administration.

Primitive masters and knights held influence over immense domains, and the serfs, who worked the land, were bound to their rulers. This progressive construction would characterize European culture for a really long time.

1.1 The Early Uses of Iron

The historical backdrop of human innovative progression is set apart by critical achievements, and the authority of iron positions among the most extraordinary. Iron, plentiful in nature and somewhat simple to remove, assumed a significant part in profoundly shaping old social orders and civilizations. Its initial purposes denoted a significant change in the capacities and potential outcomes of human networks.

The tale of iron starts with its revelation and the progress from the Bronze Age to the Iron Age, a shift that would end up being a basic defining moment in mankind's set of experiences. Prior to digging into the purposes of iron, understanding the starting points and qualities of this momentous metal is significant.

Iron is a compound component with the image Fe and nuclear number 26. It is perhaps of the most bountiful component on The planet, making up a huge piece of the planet's center. Iron mineral, normally as hematite or magnetite, is tracked down in different areas all over the planet, giving an open wellspring of the metal.

The Iron Age, which supplanted the Bronze Age, denoted a period when social orders started to involve iron for different purposes. Dissimilar to bronze, which is a compound of copper and tin, iron is an unadulterated component. The change from bronze to press was driven by the accessibility of iron metal and the information on the most proficient method to remove and control it.

One of the earliest and most significant strategies created by antiquated human advancements was refining. Refining is the most common way of separating metal from its mineral by warming the metal to high temperatures within the sight of a diminishing specialist. On account of iron, the diminishing specialist is normally charcoal. This cycle yields iron in a more moldable structure, known as fashioned iron.

The Hittites, an old Anatolian individuals who lived around 1600 BCE, are accepted to have been among quick to dominate ironworking. Their insight into refining and manufacturing iron permitted them to make predominant weapons and devices. This mechanical benefit gave them an impressive edge in fighting and added to the extension of their realm.

In India, iron refining had likely started around 1800 BCE, making it one of the early focuses of iron creation. Iron antiques from this period have been found in the locale, bearing witness to the old Indians' capability in ironworking. The iron-rich Deccan Level, specifically, assumed a critical part in this early iron industry.

The spread of ironworking innovation was not bound to a solitary district. As information on iron purifying and producing strategies spread, different old civic establishments started to integrate iron into their day to day routines. The change from bronze to press was continuous, with the two metals coinciding for quite a while as social orders adjusted to the additional opportunities presented by iron.

One of the main parts of iron's initial use was its importance in fighting. Iron weapons and defensive layer offered unmistakable benefits over their bronze partners. Iron is more diligently and more strong than bronze, making it prevalent for creating weapons that could endure the afflictions of battle. The reception of iron weapons, like swords, lances, and pointed stones, significantly affected the advancement of military innovation.

In the Close to East, the Assyrians, a strong domain focused in Mesopotamia, utilized iron weaponry during the primary thousand years BCE. The unrivaled strength and solidness of iron permitted them to vanquish and keep up with huge domains. The Assyrians' iron weapons were vital to their tactical strength and their capacity to fabricate and safeguard their domain.

The Iron Age additionally saw the rise of iron chariots, which were intensely protected and furnished with iron sickles appended to the wheels. These impressive conflict machines offered a significant benefit in fight and were utilized by different old civic establishments, including the Egyptians and the Hittites.

Iron's effect on fighting was not restricted to weaponry alone. The development of fortresses and attack motors additionally profited from the expanded accessibility of iron. Iron devices and gear empowered the development of stronger protective designs, while the creation of iron shots and battering rams worked with attack fighting.

The benefits of iron reached out to the domain of agribusiness too. Iron furrows and other cultivating apparatuses altogether worked on rural productivity. The capacity to work the dirt all the more really considered expanded food creation, which, thusly, upheld bigger populaces and the development of old urban areas.

Iron sickles and grass cutters, intended for gathering crops, were more strong and proficient than their bronze partners. This advancement added to the capacity of rural social orders to create excesses and, subsequently, participate in exchange and specialization of work. Iron instruments were a critical consider the extension of farming boondocks and the improvement of coordinated rural frameworks.

As social orders embraced ironworking, the interest for iron extended past military and agrarian applications. Iron was progressively utilized in daily existence. Iron cookware, for instance, became ordinary in families, and its solidness and intensity holding properties made it profoundly functional for preparing and food arrangement.

The antiquated Greeks, who were known for their high level information and advancements, made huge commitments to the improvement of iron innovation. The Greeks developed existing iron purifying methods, upgrading the nature of iron delivered. They likewise assumed an essential part in the scattering of ironworking information all through the Mediterranean locale.

The coming of the Roman Realm saw a further refinement of ironworking strategies. The Romans' broad street organizations and foundation projects required gigantic amounts of iron. They constructed water systems, spans, and amazing designs utilizing iron apparatuses and executes. Roman specialists planned water haggles, which were fueled by iron pinion wheels and shafts, to work with different modern cycles.

The utilization of iron in development considered the making of getting through structural wonders, like the Roman reservoir conduits. The Pont du Gard, an old Roman water passage in southern France, is a demonstration of the designing capacities of the time. Developed with exactly cut and fitted stone blocks kept intact by iron clips, it stays an image of Roman designing ability.

The Romans likewise utilized iron in their tactical designing, making noteworthy attack motors and strongholds. Iron nails, screws, and latches kept intact the parts of these huge machines. Ballistae, a kind of huge crossbow utilized for attack fighting, highlighted iron parts that considered more prominent power and precision.

The meaning of iron stretched out to transportation too. The advancement of iron-tipped nails and pony drawn carriages altered travel and exchange. Iron nails, specifically, supplanted wooden stakes in the development of boats, guaranteeing the strength and safety of vessels.

The far reaching utilization of iron nails took into account the development of bigger boats, equipped for conveying heavier freight and more travelers. This development assumed a critical part in the extension of sea shipping lanes and the trading of merchandise and thoughts across the Mediterranean and then some.

The developing accessibility of iron, combined with headways in ironworking strategies, prompted the creation of iron coins. These normalized units of money worked with exchange and financial exchanges, as they were effectively unmistakable and esteemed. The utilization of iron coins became boundless in numerous old social orders, giving a mode of trade that advanced financial solidness and development.

In old China, iron assumed a major part in the improvement of weaponry and devices. The Chinese found that by adding carbon to press during the purifying system, they could make a lot harder material: cast iron. Project iron, with its expanded hardness and toughness, was great for the creation of sharp-edged weapons and apparatuses.

The Chinese were among quick to project iron into molds to make many-sided and exceptionally improving items, a method known as lost-wax projecting. This interaction permitted them to deliver a wide cluster of things, including resplendent bronze mirrors, strict puppets, and utilitarian devices. The utilization of solid metal extended to incorporate weapons and horticultural carries out as well as perplexing masterpieces.

Iron was likewise indispensable to the improvement of the Chinese furrow, which was fundamentally more proficient than prior wooden furrows. The expanded agrarian efficiency coming about because of iron furrows added to populace development and the extension of cultivating locales.

Chinese advances in iron creation and innovation were communicated to adjoining areas, especially through exchange along the Silk Street. The spread of ironworking information significantly affected Focal Asia and the Center East, prompting developments in iron apparatuses and weaponry in these districts also.

The Center East, as a junction of societies and developments, assumed a critical part in the spread of ironworking methods. The Abbasid Caliphate, situated in Baghdad during the Islamic Brilliant Age, was a center point of logical and mechanical development. The Islamic world based upon before information on iron refining and metalworking procedures, and this information was additionally evolved and communicated to Europe.

By the early middle age period, European ironworking was advancing, with the development of the waterwheel-controlled bloomery heater. This innovation considered the effective refining of iron metal and the creation of fashioned iron. The expanded creation limit upheld the advancement of archaic Europe, from horticultural upgrades to growing metropolitan habitats.

The Viking Age, which crossed from the late eighth 100 years to the eleventh hundred years, saw broad utilization of iron in the development of longships, weapons, and apparatuses.

1.2 Ancient Iron-Making Techniques

The turn of events and dominance of iron-production methods in old times denoted a vital point in mankind's set of experiences. The capacity to concentrate, smelt, and control iron had significant ramifications for innovation, fighting, and cultural progression. The excursion to outfit this significant asset and open its possible traversed hundreds of years and achieved noteworthy advancements in metallurgy.

Iron, as a characteristic component, has been available on Earth for billions of years. It is perhaps of the most plentiful component in the World's outside layer, and its broad accessibility made it an appealing asset for early human social orders. Nonetheless, the change from perceiving iron as a valuable material to effectively removing and working with it was a complicated and progressive cycle.

Perhaps of the earliest test in iron creation was the need to arrive at the high temperatures expected for purifying. Iron mineral, in its normal state, is regularly as iron oxides, like hematite or magnetite. Refining, the most common way of extricating metal from mineral, includes warming the mineral to high temperatures within the sight of a lessening specialist, like charcoal, to diminish the iron oxides to unadulterated iron.

Bronze, a copper-tin composite, had been the essential metal utilized in old times, and its softening point was fundamentally lower than that of iron. Purifying iron required temperatures that surpassed 1,500 degrees Celsius (2,732 degrees Fahrenheit), which were a long ways past the capacities of early heaters and manufactures.

One of the primary iron-production strategies created was the bloomery heater. Bloomery heaters were little, basic designs used to smelt iron. The expression "bloomery" is gotten from "blossom," which alludes to a mass of iron that structures during the purifying system. These heaters were fabricated utilizing mud or stone and were somewhat wasteful as far as temperature control.

The bloomery heater worked on the guideline of decrease, where iron metal was warmed within the sight of charcoal. The carbon in the charcoal decreased the iron oxide to create metallic iron. Notwithstanding, the cycle likewise delivered slag, a side-effect that comprised of pollutions and non-metallic minerals.

The bloomery heater had its impediments. Its little size confined the amount of iron that could be created, and the absence of temperature control implied that the subsequent iron was frequently not completely sanitized. The iron delivered in bloomery heaters was normally as created iron, which had a high carbon content and was not generally so hard as steel.

The change from bronze to press denoted the start of the Iron Age. It was a time of trial and error and development as social orders looked to dominate the procedures expected for proficient iron creation. The Hittites, an old Anatolian progress, are accepted to have been among quick to foster the information and abilities vital for ironworking. Their utilization of iron weaponry gave them a critical military benefit, and the information on iron refining spread all through the old world.

In antiquated India, iron purifying likely started around 1800 BCE, making the Indian subcontinent one of the early focuses of iron creation. Iron curios from this period have been found, confirming the capability of antiquated Indians in ironworking. The iron-rich Deccan Level assumed a critical part in this early iron industry.

The spread of ironworking innovation was not restricted to a solitary locale. As information on iron refining and producing strategies spread, different antiquated civilizations started to integrate iron into their day to day routines.

The progress from bronze to press was steady, with the two metals coinciding for quite a while as social orders adjusted to the additional opportunities presented by iron.

One of the main parts of early iron creation was its importance in fighting. Iron weaponry, with its prevalent strength and toughness, offered particular benefits over bronze weapons. The boundless reception of iron weapons, like blades, lances, and sharpened stones, significantly affected the advancement of military innovation.

The Assyrians, an old realm focused in Mesopotamia, utilized iron weaponry during the primary thousand years BCE. Their dominance of iron innovation gave them a tactical benefit and added to the extension of their domain. Iron weapons, including blades and reinforcement, assumed a focal part in their tactical strength and their capacity to construct and shield their huge regions.

The Iron Age likewise saw the advancement of iron chariots, vigorously shielded vehicles outfitted with iron grass cutters connected to the wheels. These imposing conflict machines offered an impressive benefit in fight and were utilized by different old civilizations, including the Egyptians and the Hittites.

Iron's effect on fighting was not restricted to weaponry alone. The development of fortresses and attack motors additionally profited from the expanded accessibility of iron. Iron apparatuses and hardware empowered the development of stronger protective designs, while the creation of iron shots and battering rams worked with attack fighting.

The benefits of iron stretched out to the domain of horticulture too. Iron furrows and other cultivating instruments altogether worked on horticultural productivity. The capacity to work the dirt all the more successfully considered expanded food creation, which, thus, upheld bigger populaces and the development of antiquated urban communities.

Iron sickles and grass cutters, intended for gathering crops, were more strong and effective than their bronze partners. This advancement added to the capacity of farming social orders to create excesses and, thusly, participate in exchange and specialization of work. Iron devices were a vital consider the extension of horticultural boondocks and the improvement of coordinated farming frameworks.

As social orders embraced ironworking, the interest for iron extended past military and rural applications. Iron was progressively utilized in day to day existence. Iron cookware, for instance, became ordinary in families, and its sturdiness and intensity holding properties made it exceptionally functional for preparing and food arrangement.

The old Greeks, known for their high level information and advancements, made huge commitments to the improvement of iron innovation. The Greeks refined existing iron purifying procedures, improving the nature of iron delivered. They likewise assumed a pivotal part in the spread of ironworking information all through the Mediterranean district.

The coming of the Roman Realm saw a further refinement of ironworking methods. The Romans' broad street organizations and framework projects required monstrous amounts of iron. They fabricated reservoir conduits, spans, and amazing designs utilizing iron apparatuses and carries out. Roman specialists planned water haggles, which were controlled by iron cog wheels and shafts, to work with different modern cycles.

The utilization of iron in development considered the making of getting through engineering wonders, like the Roman reservoir conduits. The Pont du Gard, an old Roman water passage in southern France, is a demonstration of the designing capacities of the time. Developed with unequivocally cut and fitted stone blocks kept intact by iron clasps, it stays an image of Roman designing ability.

The Romans likewise utilized iron in their tactical designing, making great attack motors and fortresses. Iron nails, screws, and clasp kept intact the parts of these huge machines. Ballistae, a kind of huge crossbow utilized for attack fighting, included iron parts that took into consideration more prominent power and exactness.

The meaning of iron stretched out to transportation also. The advancement of iron-tipped nails and pony drawn carriages upset travel and exchange. Iron nails, specifically, supplanted wooden stakes in the development of boats, guaranteeing the sturdiness and fitness for sailing of vessels.

The boundless utilization of iron nails considered the development of bigger boats, fit for conveying heavier freight and more travelers. This development assumed a crucial part in the extension of oceanic shipping lanes and the trading of merchandise and thoughts across the Mediterranean and then some.

The developing accessibility of iron, combined with progressions in ironworking strategies, prompted the creation of iron coins. These normalized units of money worked with exchange and financial exchanges, as they were effectively conspicuous and esteemed. The utilization of iron coins became broad in numerous old social orders, giving a mode of trade that advanced monetary solidness and development.

In old China, iron assumed a central part in the improvement of weaponry and devices. The Chinese found that by adding carbon to press during the purifying system, they could make a lot harder material: cast iron. Project iron, with its expanded hardness and sturdiness, was great for the creation of sharp-edged weapons and apparatuses.

The Chinese were among quick to project iron into molds to make complex and profoundly enriching objects, a procedure known as lost-wax projecting. This cycle permitted them to create a wide cluster of things, including fancy bronze mirrors, strict dolls, and utilitarian instruments. The utilization of solid metal extended to incorporate weapons and farming executes as well as complex show-stoppers.

Iron was additionally necessary to the advancement of the Chinese furrow, which was altogether more proficient than prior wooden furrows. The expanded rural efficiency coming about because of iron furrows added to populace development and the extension of cultivating districts.

Chinese advances in iron creation and innovation were sent to adjoining areas, especially through exchange along the Silk Street. The spread of ironworking information significantly affected Focal Asia and the Center East, prompting developments in iron apparatuses and weaponry in these districts also.

1.3 Iron in Warfare and Society

The job of iron in fighting and society has been a characterizing component of mankind's set of experiences. The authority of ironworking procedures and the broad utilization of iron had significant ramifications for military system, agribusiness, framework advancement, and financial development. The change from the Bronze Age to the Iron Age denoted an essential second in the development of human progress, as iron offered unrivaled strength, solidness, and flexibility contrasted with before metals.

The reception of iron weaponry assumed a focal part in molding the course of fighting in old times. Iron's benefits over bronze, the essential metal utilized during the Bronze Age, were promptly clear. Iron is more enthusiastically and more sturdy than bronze, making it ideal for creating weapons that could endure the afflictions of battle.

The Hittites, an old Anatolian individuals who lived around 1600 BCE, are accepted to have been among quick to dominate ironworking. Their insight into refining and fashioning iron permitted them to make unrivaled weapons and devices. This mechanical benefit gave them an impressive edge in fighting and added to the extension of their domain.

In India, iron purifying likely started around 1800 BCE, making it one of the early focuses of iron creation. Iron curios from this period have been found in the district, authenticating the old Indians' capability in ironworking. The iron-rich Deccan Level, specifically, assumed a huge part in this early iron industry.

The spread of ironworking innovation was not restricted to a solitary locale. As information on iron refining and manufacturing procedures spread, different old developments started to integrate iron into their day to day routines. The progress from bronze to press was slow, with the two metals coinciding for quite a while as social orders adjusted to the additional opportunities presented by iron.

One of the main parts of iron's initial use was its effect on military innovation. Iron weapons, including blades, lances, and sharpened stones, gave particular benefits in fight. The more noteworthy hardness and toughness of iron considered the development of additional successful and imposing weapons.

The Assyrians, an old domain focused in Mesopotamia, utilized iron weaponry during the primary thousand years BCE. The predominant strength and sturdiness of iron gave them a significant military benefit, empowering them to vanquish and keep up with tremendous regions. Iron weapons, covering, and strongholds were key to their tactical strength and their capacity to fabricate and guard their realm.

The Iron Age likewise saw the advancement of iron chariots, vigorously defensively covered vehicles outfitted with iron sickles joined to the wheels. These imposing conflict machines offered an impressive benefit in fight and were utilized by different old developments, including the Egyptians and the Hittites.

Iron's effect on fighting stretched out past weaponry. The development of strongholds and attack motors additionally profited from the expanded accessibility of iron. Iron devices and hardware empowered the formation of stronger protective designs, while the development of iron shots and battering rams worked with attack fighting.

In old China, iron assumed a principal part in the improvement of weaponry and devices. Chinese metallurgists found that by adding carbon to press during the refining system, they could make a lot harder material: cast iron. Project iron, with its expanded hardness and strength, was great for the development of sharp-edged weapons and instruments.

The Chinese were among quick to project iron into molds to make complex and exceptionally beautiful items, a method known as lost-wax projecting. This interaction permitted them to deliver a wide cluster of things, including elaborate bronze mirrors, strict puppets, and utilitarian instruments. The utilization of solid metal extended to incorporate weapons and agrarian carries out as well as perplexing masterpieces.

Iron was likewise necessary to the improvement of the Chinese furrow, which was essentially more proficient than prior wooden furrows. The expanded rural efficiency coming about because of iron furrows added to populace development and the extension of cultivating districts.

Chinese advances in iron creation and innovation were sent to adjoining areas, especially through exchange along the Silk Street. The spread of ironworking information significantly affected Focal Asia and the Center East, prompting developments in iron apparatuses and weaponry in these districts also.

The Center East, as a junction of societies and civic establishments, assumed a critical part in the scattering of ironworking strategies. The Abbasid Caliphate, situated in Baghdad during the Islamic Brilliant Age, was a center of logical and mechanical development. The Islamic world based upon before information on iron refining and metalworking procedures, further creating and sending this information to Europe.

By the early archaic period, European ironworking was developing, with the rise of the waterwheel-fueled bloomery heater. This innovation took into account the effective refining of iron metal and the development of fashioned iron. The expanded creation limit upheld the advancement of archaic Europe, from rural enhancements to extending metropolitan focuses.

The Viking Age, which crossed from the late eighth hundred years to the eleventh hundred years, saw broad utilization of iron in the development of longships, weapons, and apparatuses. Iron tomahawks and blades, known for their quality and craftsmanship, were profoundly pursued by Viking champions and brokers.

The job of iron in middle age Europe went past viable applications. The iron crossbow, a strong and innovatively progressed weapon, turned into an image of middle age fighting. It was an early forerunner to the more refined guns that would ultimately alter fighting.

The change from the middle age to the Renaissance time frame got further progressions ironworking. The improvement of impact heaters considered the creation of solid metal for a bigger scope. Project iron was utilized to make many-sided design subtleties and embellishments for places of worship, royal residences, and different designs.

The Renaissance likewise saw the introduction of the advanced steel industry. Steel, a composite of iron and carbon, offers much more noteworthy strength and flexibility than cast iron or created iron. The most common way of changing over iron into steel includes cautiously controlling the carbon content through strategies like cementation or rankle steel creation.

Steel quickly turned into the material of decision for creating weapons and defensive layer. It gave the best mix of solidarity, strength, and sharpness. The improvement of steel took into consideration the formation of the notorious longswords and blades utilized by European knights and fighters.

The presentation of explosive and the improvement of guns denoted a critical defining moment throughout the entire existence of weaponry. Guns depended on the utilization of iron and steel for their development, and this change had sweeping ramifications for military strategies and methodology. The period of explosive fighting had shown up.

The utilization of iron additionally added to the advancement of apparatus and modern cycles. Iron cog wheels, shafts, and parts assumed a basic part in the Modern Upheaval, which started in the late eighteenth 100 years. This time of significant financial and mechanical change saw the boundless reception of steam motors, automated creation, and the development of production line based assembling.

The development of rail lines and trains, fueled by iron and later steel, reformed transportation and exchange. Iron extensions and passages considered the development of railroad organizations, associating far off areas and working with the development of products and individuals on an extraordinary scale.

The late nineteenth and mid twentieth hundreds of years saw further developments in iron and steel innovation. The improvement of the Bessemer interaction, a technique for efficiently manufacturing steel, upset the steel business by decisively lessening the expense of steel creation. This headway made steel more open for different applications, from development to modern hardware.

Iron's job in development arrived at new levels with the Eiffel Pinnacle, a wonder of designing that was fabricated utilizing in excess of 18,000 individual iron pieces. The pinnacle's development displayed the flexibility and strength of iron, and it turned into a famous image of the Modern Age.

As the twentieth century advanced, the car business started to depend intensely on iron and steel for the development of vehicles, trucks, and apparatus. Iron amalgams gave the fundamental strength and sturdiness for motor parts and skeleton development.

During The Second Great War, iron and steel creation turned into a foundation of the conflict exertion. These materials were fundamental for the production of tanks, boats, airplane, and weaponry. The capacity to efficiently manufacture iron and steel assumed a fundamental part in the conflict's result.

The post-war time frame saw the development of the worldwide steel industry, with an emphasis on large scale manufacturing and mechanical progressions. Steel turned into a critical material in the development of high rises, scaffolds, and framework, reshaping metropolitan scenes all over the planet.

Iron's effect on society stretched out past fighting and development. Its job in agribusiness was groundbreaking. Iron furrows and other cultivating instruments essentially worked on rural effectiveness. The capacity to work the dirt all the more really considered expanded food creation, which, thusly, upheld bigger populaces and the development of antiquated urban communities.

Iron sickles and grass cutters, intended for reaping crops, were more strong and effective than their bronze partners. This advancement added to the capacity of horticultural social orders to deliver excesses and, subsequently, participate in exchange and specialization of work. Iron devices were a vital calculate the development .

Chapter 2

Iron and the Industrial Revolution

The Modern Upheaval was a groundbreaking period in mankind's set of experiences that reshaped social orders, economies, and advancements. At its center, the transformation was portrayed by the quick and significant shift from agrarian and craftsmanship based economies to modern and motorized ones. Key to this change was the crucial pretended by iron, both as a material and as an impetus for development.

The Modern Unrest had its starting points in England during the late eighteen hundred years and consequently spread to different pieces of Europe and the US. This epochal shift was driven by a combination of elements, including mechanical progressions, changes in friendly and monetary designs, and a developing requirement for expanded creation and proficiency.

The development of the iron and steel industry was a characterizing element of the Modern Upset. The use of iron and steel for hardware, foundation, and transportation frameworks significantly affected the advancement of present day modern social orders. This exposition investigates the vital job of iron and steel in the Modern Unrest, following the advancement of iron creation and different applications supported the emotional cultural changes of the time.

The Iron Business in the Pre-Modern Period

Before the Modern Upset, iron creation was basically founded on conventional procedures, including the utilization of bloomery heaters. These heaters were little, dirt or stone designs used to smelt iron mineral and concentrate the metal. The bloomery heater, while effective for creating little amounts of iron, was restricted in its ability, and the subsequent iron was commonly of lower quality, known as fashioned iron.

Created iron, described by its flexibility and moderately low carbon content, was appropriate for specific applications, for example, blacksmithing and manufacturing, however it coming up short on strength and solidness expected for huge scope modern hardware and development. Subsequently, early modern endeavors confronted imperatives because of the impediments of customary iron creation.

The advancement of the iron business in the pre-Modern Unrest time frame was set apart by steady upgrades in iron creation methods. In the seventeenth and eighteenth hundreds of years, developments, for example, coke purifying and the utilization of water-fueled hammers started to lay the foundation for more proficient iron creation. These developments, in any case, were as yet a long ways from the progressive changes that would happen during the Modern Upset.

The Approach of the Impact Heater

The genuine leap forward in iron creation accompanied the reception of the impact heater, an innovation that fundamentally expanded iron result and worked on its quality. The impact heater, which had its starting points in early current Europe, was described by its huge size, high limit, and the utilization of constrained air (impacts) to accomplish higher temperatures.

The impact heater worked on an unexpected rule in comparison to the bloomery heater. It could smelt iron mineral at temperatures surpassing 1,500 degrees Celsius (2,732 degrees Fahrenheit), a basic calculate creating cast iron, a material with a higher carbon content and more noteworthy strength than fashioned iron.

The most common way of purifying iron in an impact heater included a progression of steps:

Accusing the heater of iron mineral, coke (a refined type of coal), and limestone.

Lighting the heater and permitting the materials to warm up.

Infusing constrained air, commonly given by a waterwheel-driven howls, into the heater to accomplish higher temperatures.

The iron mineral would lessen into liquid iron, with pollutions (known as slag) settling at the base.

Tapping the liquid iron and projecting it into molds, bringing about pig iron or cast iron.

The capacity to deliver cast iron in enormous amounts was a critical headway in metallurgy. Project iron had a higher carbon content, which made it harder and more fragile than created iron yet in addition reasonable for different modern applications. It was more grounded and more sturdy than prior materials, pursuing it an optimal decision for building apparatus and designs. Besides, cast iron's adaptability and formability took into consideration the production of multifaceted and lavish structural plans.

The reception of the shoot heater in iron creation was instrumental in the advancement of the iron and steel industry, however it likewise had more extensive cultural ramifications. The expanded accessibility of iron and solid metal considered the making of many items and framework that became integral to the Modern Insurgency.

The Multiplication of Iron and Steel in Apparatus

One of the main effects of the Modern Upheaval was the multiplication of apparatus. The coming of the steam motor, a progressive development by its own doing, was firmly interwoven with the utilization of iron and steel. Iron was a crucial material in the development of steam motors, which controlled manufacturing plants, trains, ships, and an extensive variety of modern hardware.

The steam motor, at first created by specialists, for example, Thomas Newcomen and James Watt, depended on steam to produce mechanical power. The cylinder and chamber, key parts of the steam motor, were built utilizing iron. Iron's solidarity and sturdiness permitted these motors to work effectively and dependably, making them integral to the modern scene.

The improvement of steam motors denoted the start of another time in transportation and assembling. Processing plants could now be found away from water sources, as steam motors gave a convenient and flexible wellspring of force. This change in assembling and creation established the groundwork for the development of modern urban areas and the processing plant framework.

The utilization of iron in hardware reached out to different other modern areas. Iron pinion wheels and parts empowered the production of multifaceted apparatus for material factories, paper plants, and mining activities. These headways expanded creation limit and proficiency, working with the development of various ventures.

The Iron Scaffold, developed in 1779 by Abraham Darby III in Coalbrookdale, Britain, is a commended illustration of iron's likely in designing and engineering. It was the main solid metal scaffold on the planet and exhibited the potential outcomes of iron in development. The Iron Extension changed transportation as well as displayed the style of given iron a role as a structure material.

Iron and steel were fundamental in the development of rail route frameworks that reformed transportation during the Modern Unrest. The improvement of trains, railroad tracks, and scaffolds depended intensely on iron and steel. Project iron was utilized for different parts of the rail route, while steel rails turned into the norm for tracks because of their solidarity and toughness.

The broad reception of rail routes significantly affected society, empowering quicker and more dependable transportation of merchandise and individuals. It prompted the making of broad railroad organizations, associating urban communities and locales and working with the development of unrefined components and completed items.

Iron and Steel in Foundation and Engineering

The expanded accessibility of iron and steel likewise assumed a crucial part in the improvement of foundation and engineering. Enormous iron designs, like extensions and structures, became images of the Modern Age.

Iron extensions, specifically, addressed a victory of designing and plan. The utilization of solid metal took into consideration unpredictable grid structures that were not practical with different materials. The Britannia Scaffold, finished in 1850 in Grains, and the Tay Extension, finished in 1878 in Scotland, are prominent instances of iron rail route spans that displayed the capability of iron in development.

The Eiffel Pinnacle, finished in 1889 for the Work Universelle in Paris, is a widely popular image of iron's capacities in engineering. Planned by engineer Gustave Eiffel, the pinnacle's many-sided cross section structure was built utilizing north of 18,000 individual iron pieces. Its development shown the strength and flexibility of iron as a structure material and stays a getting through symbol of the Modern Transformation.

Iron's utilization in framework stretched out to the development of water systems, production lines, stockrooms, and other modern and public structures. Its solidarity and formability considered the making of bigger and more strong designs. Iron segments and pillars upheld multi-story structures, denoting a takeoff from prior compositional styles.

The improvement of the Bessemer cycle, a strategy for efficiently manufacturing steel, further extended the opportunities for design and development. Steel, with its higher strength and solidness contrasted with cast iron, turned into a favored material for primary components in structures and scaffolds.

The Gem Royal residence, built in London for the Incomparable Display of 1851, was a noteworthy illustration of the utilization of iron and glass in engineering. The construction included a cast iron and glass structure, permitting regular light to enlighten the inside. This inventive plan, made conceivable by the utilization of iron and glass, addressed a takeoff from conventional structure techniques.

The effect of iron and steel in development was not restricted to enormous scope projects. The utilization of iron nails, screws, and clasp reformed carpentry and carpentry, making it more straightforward to join and develop wooden designs. The accessibility of normalized iron parts, for example, nails and screws, worked with development processes and added to the development of the development business.

Iron and the Extension of Urbanization

The extension of metropolitan focuses and the development of modern urban communities were immediate results of the Modern Unrest. Iron assumed a focal part in this change. The development of production lines, factories, and modern offices required huge amounts of iron and steel. Iron pillars and sections were utilized .

2.1 Iron as the Backbone of the Industrial Revolution

The Modern Unrest, a groundbreaking period that reshaped social orders, economies, and innovations, was in a general sense supported by iron. The combination of variables, including mechanical progressions, changes in friendly and financial designs, and the rising requirement for higher creation and effectiveness, characterized this epochal shift. The critical pretended by iron, both as a material and as an impetus for development, was key to this change.

The Modern Unrest had its beginnings in England during the late eighteenth hundred years and consequently spread to different pieces of Europe and the US. It was set apart by the quick and significant shift from agrarian and craftsmanship based economies to modern and motorized ones. The development of the iron and steel industry was a characterizing element of this transformation. The usage of iron and steel for hardware, framework, and transportation frameworks significantly affected the improvement of present day modern social orders. This paper investigates the fundamental job of iron and steel in the Modern Upset, following the advancement of iron creation and various applications supported the sensational cultural changes of the period.

The Iron Business in the Pre-Modern Period

Before the Modern Upset, iron creation was fundamentally founded on conventional methods, including the utilization of bloomery heaters. These heaters were little, mud or stone designs used to smelt iron mineral and concentrate the metal. The bloomery heater, while proficient for delivering little amounts of iron, was restricted in its ability, and the subsequent iron was commonly of lower quality, known as fashioned iron.

Fashioned iron, portrayed by its flexibility and generally low carbon content, was appropriate for specific applications, for example, blacksmithing and producing, yet it missing the mark on strength and sturdiness expected for huge scope modern apparatus and development. Thus, early modern endeavors confronted requirements because of the constraints of conventional iron creation.

The advancement of the iron business in the pre-Modern Upheaval time frame was set apart by steady upgrades in iron creation methods. In the seventeenth and eighteenth hundreds of years, advancements, for example, coke purifying and the utilization of water-controlled hammers started to lay the foundation for more effective iron creation. These developments, nonetheless, were as yet a long ways from the progressive changes that would happen during the Modern Upset.

The Coming of the Impact Heater

The genuine forward leap in iron creation accompanied the reception of the impact heater, an innovation that essentially expanded iron result and worked on its quality. The impact heater, which had its starting points in early current Europe, was portrayed by its huge size, high limit, and the utilization of constrained air (impacts) to accomplish higher temperatures.

The impact heater worked on an unexpected guideline in comparison to the bloomery heater. It could smelt iron mineral at temperatures surpassing 1,500 degrees Celsius (2,732 degrees Fahrenheit), a basic figure creating cast iron, a material with a higher carbon content and more prominent strength than fashioned iron.

The most common way of purifying iron in an impact heater included a progression of steps:

Accusing the heater of iron mineral, coke (a refined type of coal), and limestone.

Lighting the heater and permitting the materials to warm up.

Infusing constrained air, regularly given by a waterwheel-driven roars, into the heater to accomplish higher temperatures.

The iron mineral would diminish into liquid iron, with contaminations (known as slag) settling at the base.

Tapping the liquid iron and projecting it into molds, bringing about pig iron or cast iron.

The capacity to deliver cast iron in huge amounts was a critical headway in metallurgy. Project iron had a higher carbon content, which made it harder and more fragile than fashioned iron yet additionally reasonable for different modern applications. It was more grounded and more sturdy than prior materials, going with it an optimal decision for building hardware and designs. Besides, cast iron's flexibility and formability took into consideration the production of mind boggling and luxurious engineering plans.

The reception of the shoot heater in iron creation was instrumental in the advancement of the iron and steel industry, yet it likewise had more extensive cultural ramifications. The expanded accessibility of iron and solid metal considered the formation of a large number of items and foundation that became fundamental to the Modern Upheaval.

The Expansion of Iron and Steel in Apparatus

One of the main effects of the Modern Unrest was the multiplication of hardware. The coming of the steam motor, a progressive creation by its own doing, was firmly interwoven with the utilization of iron and steel. Iron was a basic material in the development of steam motors, which controlled processing plants, trains, ships, and an extensive variety of modern hardware.

The steam motor, at first created by specialists, for example, Thomas Newcomen and James Watt, depended on steam to produce mechanical power. The cylinder and chamber, key parts of the steam motor, were developed utilizing iron. Iron's solidarity and strength permitted these motors to work proficiently and dependably, making them key to the modern scene.

The improvement of steam motors denoted the start of another time in transportation and assembling. Plants could now be found away from water sources, as steam motors gave a convenient and flexible wellspring of force. This change in assembling and creation established the groundwork for the development of modern urban communities and the processing plant framework.

The utilization of iron in hardware stretched out to different other modern areas. Iron pinion wheels and parts empowered the production of perplexing apparatus for material plants, paper factories, and mining tasks. These headways expanded creation limit and proficiency, working with the development of various enterprises.

The Iron Extension, developed in 1779 by Abraham Darby III in Coalbrookdale, Britain, is a praised illustration of iron's possible in designing and engineering. It was the main solid metal scaffold on the planet and exhibited the conceivable outcomes of iron in development. The Iron Extension upset transportation as well as displayed the style of given iron a role as a structure material.

Iron and steel were fundamental in the development of rail route frameworks that upset transportation during the Modern Transformation. The advancement of trains, rail route tracks, and extensions depended vigorously on iron and steel. Project iron was utilized for different parts of the railroad, while steel rails turned into the norm for tracks because of their solidarity and strength.

The far reaching reception of rail lines significantly affected society, empowering quicker and more dependable transportation of products and individuals. It prompted the making of broad rail route organizations, interfacing urban communities and areas and working with the development of unrefined components and completed items.

Iron and Steel in Foundation and Design

The expanded accessibility of iron and steel likewise assumed a crucial part in the improvement of framework and design. Huge iron designs, like scaffolds and structures, became images of the Modern Age.

Iron scaffolds, specifically, addressed a victory of designing and plan. The utilization of solid metal took into consideration many-sided cross section structures that were not attainable with different materials. The Britannia Scaffold, finished in 1850 in Grains, and the Tay Extension, finished in 1878 in Scotland, are outstanding instances of iron rail line spans that displayed the capability of iron in development.

The Eiffel Pinnacle, finished in 1889 for the Work Universelle in Paris, is a widely popular image of iron's capacities in engineering. Planned by engineer Gustave Eiffel, the pinnacle's complex cross section structure was developed utilizing north of 18,000 individual iron pieces. Its development shown the strength and flexibility of iron as a structure material and stays a getting through symbol of the Modern Upheaval.

Iron's utilization in framework stretched out to the development of water systems, production lines, stockrooms, and other modern and public structures. Its solidarity and formability took into account the production of bigger and more strong designs. Iron segments and bars upheld multi-story structures, denoting a takeoff from prior design styles.

The improvement of the Bessemer interaction, a technique for efficiently manufacturing steel, further extended the opportunities for engineering and development. Steel, with its higher strength and solidness contrasted with cast iron, turned into a favored material for primary components in structures and scaffolds.

The Precious stone Castle, built in London for the Incomparable Display of 1851, was a noteworthy illustration of the utilization of iron and glass in design. The design included a cast iron and glass structure, permitting normal light to enlighten the inside. This inventive plan, made conceivable by the utilization of iron and glass, addressed a takeoff from customary structure strategies.

The effect of iron and steel in development was not restricted to enormous scope projects. The utilization of iron nails, screws, and latches reformed carpentry and carpentry, making it simpler to join and build wooden designs. The accessibility of normalized iron parts, for example, nails and screws, worked with development processes and added to the development of the development business.

Iron and the Development of Urbanization

The development of metropolitan focuses and the development of modern urban communities were immediate results of the Modern Unrest. Iron assumed a focal part in this change. The development of processing plants, factories, and modern offices required enormous amounts of iron and steel. Iron shafts and sections were utilized to make multi-story manufacturing plants .

2.2 Steam Engines and Iron Machinery

The Modern Upset, an extraordinary period in mankind's set of experiences, was driven by a combination of variables that reshaped social orders, economies, and advancements. Integral to this change was the significant job of steam motors and iron apparatus, which on a very basic level modified how merchandise were delivered, shipped, and fueled. The mix of steam power and iron hardware denoted a defining moment in industrialization, preparing for present day modern social orders.

The Modern Unrest had its starting points in England during the late eighteenth hundred years before thusly spreading to different pieces of Europe and the US. This epochal shift was described by the fast and significant progress from agrarian and craftsmanship based economies to modern and motorized ones. The coming of steam motors, at first created by designers like Thomas Newcomen and James Watt, and their reconciliation with iron hardware assumed a basic part in driving this upset.

The improvement of steam motors and their application in different enterprises denoted a significant shift from customary power sources, for example, human and creature work, water wheels, and windmills, to a more dependable and flexible well-spring of mechanical power. Steam motors, powered by coal, empowered plants, transportation frameworks, and hardware to work all the more productively, speeding up creation and changing the metropolitan scene. This article investigates the necessary job of steam motors and iron hardware in the Modern Upset, following their development and the assorted applications that supported the sensational cultural changes of the time.

The Development of the Steam Motor

The steam motor, a progressive creation, filled in as the foundation of the Modern Unrest. It bridled the force of steam to produce mechanical energy and considered the motorization of different modern cycles. The improvement of steam motors was a consequence of hundreds of years of trial and error and development, finishing in crafted by Thomas Newcomen and James Watt.

The Early Steam Motors: Thomas Newcomen

The primary viable steam motor was created by Thomas Newcomen in the mid eighteenth hundred years. Known as the Newcomen motor or environmental motor, it was fundamentally used to siphon water out of mines. The motor worked on a basic guideline: steam was infused into a chamber, making it gather and make a vacuum. Environmental strain then, at that point, pushed a cylinder descending, performing mechanical work.

Newcomen's motor was a huge headway in coal mining, as it empowered further and more productive extraction of coal. Be that as it may, it was restricted in its proficiency and applications because of its intrinsic plan. The motor's shortcoming was to a great extent a consequence of the consistent warming and cooling of the chamber, bringing about energy misfortunes.

James Watt's Enhancements

The basic forward leap in steam motor innovation accompanied the upgrades made by Scottish designer James Watt during the 1760s. Watt's advancements tended to the principal impediments of the Newcomen motor, prompting huge proficiency gains.

Watt's key developments incorporated the expansion of a different condenser, which permitted the chamber to stay hot, and the presentation of a twofold acting chamber, empowering steam to perform work on both the upstroke and down-stroke. These upgrades decreased energy misfortunes and fundamentally improved the motor's proficiency.

Watt's steam motor turned into a flexible power source, as it very well may be utilized for a large number of uses past siphoning water out of mines. It fueled production lines, plants, and different sorts of hardware, making it a main thrust of industrialization. The utilization of the Watt motor in modern settings prompted the automation of creation processes and established the groundwork for the plant framework.

Watt's steam motor additionally tracked down use in transportation. It was adjusted to drive trains and steamships, changing how products and individuals were shipped over significant distances. The improvement of railroads and steam-ships assumed a critical part in the development of worldwide exchange and the development of populaces.

The Joining of Iron Apparatus

The coming of steam motors significantly affected the advancement of iron apparatus. Iron, with its solidarity, strength, and formability, was an optimal material for developing apparatus and motors. The blend of steam power and iron hardware laid the basis for the automation of various ventures.

Assembling and Material Factories

The mix of steam motors into assembling and material factories changed these enterprises. The material business, specifically, encountered an upset underway techniques. The cotton gin, turning jenny, and power loom were among the advancements that, when joined with steam power, extraordinarily expanded the effectiveness and limit of material creation.

Industrial facilities were presently ready to work for a bigger scope, and steam-driven hardware empowered the motorization of errands recently performed manually. This change in assembling rehearses denoted the introduction of the plant framework, with brought together areas for creation and a division of work.

Iron hardware, controlled by steam motors, assumed a basic part in the large scale manufacturing of materials and different products. Plants, highlighting iron bars and segments to help huge open spaces, were built to house the apparatus and labor force. The plan of these modern structures altogether affected metropolitan scenes.

Mining and Metallurgy

Steam motors and iron apparatus were likewise essential to the mining and metallurgical ventures. The use of steam power in digging considered further and more productive extraction of coal, iron metal, and different minerals. Iron siphons and motors were utilized to empty water out of mines, making it conceivable to get to beforehand blocked off assets.

The utilization of iron hardware in metallurgy was groundbreaking. The Bessemer cycle, a technique for efficiently manufacturing steel, depended on the accessibility of top notch iron mineral and coal, the two of which were made more open through steam-controlled mining tasks. The Bessemer cycle, which included blowing air through liquid iron to eliminate debasements and make steel, assumed a significant part in the modernization of industry and foundation.

Transportation and Steam Trains

The improvement of railroads, fueled by steam trains, was perhaps of the main headway in transportation during the Modern Unrest. Iron rails and steam trains considered quicker, more dependable, and proficient transportation of products and individuals.

The joining of iron apparatus into the rail route framework had extensive ramifications. Iron rails, with their solidarity and sturdiness, turned into the norm for tracks, giving a smooth and stable surface for trains to go on. Steam trains, with their strong motors, could pull weighty loads and cross significant distances, empowering the development of rail organizations.

The rail route framework associated urban areas and locales, working with the development of natural substances, completed items, and individuals. The productive transportation of merchandise brought down costs, decreased travel times, and advanced monetary development. The development of rail route networks changed the metropolitan scene, as stations, extensions, and tracks were built utilizing iron.

Farming and Steam-Fueled Hardware

Horticulture was not insusceptible to the extraordinary force of steam motors and iron apparatus. The reception of steam-fueled rural apparatus reformed cultivating works on, expanding efficiency and effectiveness. Steam farm vehicles, sifting machines, and furrowing motors were among the developments that changed the substance of horticulture.

Steam-driven furrowing motors, for instance, could work the dirt more effectively than conventional techniques, prompting expanded food creation. The capacity to deliver excesses upheld bigger populaces and the development of old urban communities. Steam-fueled sifting machines smoothed out the most common way of isolating grains from waste, diminishing the work expected for agrarian errands.

The boundless utilization of steam-controlled hardware in farming significantly affected country economies and added to the development of modern social orders. These advancements worked with the automation of rural cycles, decreasing the interest for difficult work in the cultivating area.

Urbanization and Iron Hardware

The blend of steam motors and iron hardware assumed a urgent part in the extension of urbanization during the Modern Unrest. As production lines, factories, and modern offices fueled by steam motors multiplied, modern urban communities developed, and the metropolitan scene went through huge changes.

The development of manufacturing plants required huge amounts of iron and steel. Iron shafts and sections upheld multi-story structures, empowering enormous open spaces inside modern designs. These sweeping insides were appropriate for lodging hardware, plants' backbone, and supporting effective creation processes.

Urbanization remained inseparable with industrialization, as the development of modern urban communities turned into a sign of the Modern Insurgency. The accessibility of steam power and iron hardware prodded the advancement of assembling focuses, with production lines at the center of metropolitan extension.

The presence of processing plants had broad ramifications for the segment and social designs of these metropolitan regions. Provincial populaces rushed to modern urban areas looking for work in the expanding production lines. The commitment of work in the processing plants and the appeal of higher wages drew individuals from agrarian life. Urbanization brought new open doors and difficulties, reshaping lodging, work elements, and public administrations.

2.3 Iron's Impact on Factories and Production

The Modern Transformation, a watershed period that reshaped social orders, economies, and innovations, was on a very basic level supported by iron. The assembly of elements, including mechanical headways, changes in friendly and financial designs, and the rising requirement for higher creation and productivity, characterized this epochal change. Iron, both as a material and as an impetus for development, assumed a focal part in this modern cauldron. This exposition investigates the significant effect of iron on processing plants and creation during the Modern Upset, following the advancement of iron's job in automation, and various applications supported the emotional cultural changes of the time.

Before the Modern Unrest, fabricating was prevalently completed in limited scope studios and homes, with creation restricted by difficult work and straightforward apparatuses. The progress to plants, driven by iron hardware and developments underway strategies, denoted an essential change in assembling processes. Processing plants, frequently portrayed by enormous, meticulously designed structures, unified creation and united apparatus and work on a phenomenal scale.

Iron Hardware and Motorization

Iron hardware was at the core of the processing plant framework, changing creation strategies and empowering the motorization of different ventures. The coming of steam motors, controlled by coal, assumed a urgent part in the change to motorized manufacturing plants. Steam motors, built with iron parts, gave a compact and flexible wellspring of mechanical power, permitting processing plants to be found away from water sources and making the way for large scale manufacturing.

The cotton material industry, one of the early adopters of motorization, gives a remarkable illustration of the effect of iron hardware on plants. Developments like the turning jenny, water casing, and power loom, frequently built with iron parts, upset material creation. These machines could perform assignments recently finished manually and were fueled by steam motors. Plants outfitted with these machines fundamentally expanded the productivity and limit of material creation.

The motorization of material processing plants brought about higher creation volumes and diminished work prerequisites. Laborers worked machines, and the plant framework presented a division of work, where every specialist played out a particular undertaking in the creation cycle. This division of work and the utilization of apparatus were focal components of the industrial facility framework, altogether expanding efficiency and result.

Iron hardware tracked down applications across different enterprises, including mining, metallurgy, and transportation. Steam-fueled motors, developed with iron parts, empowered more productive mining tasks, working with more profound extraction of coal, iron metal, and different minerals. The Bessemer cycle, a historic technique for efficiently manufacturing steel, depended on the accessibility of excellent iron metal and coal, both made more open through steam-fueled mining.

The utilization of iron hardware in transportation assumed a critical part in the development of rail lines. Iron rails, developed with iron of high strength and solidness, gave a steady surface to trains, and steam trains, controlled by steam motors, pulled weighty loads and navigated significant distances. The advancement of rail lines, with their iron tracks and trains, altered how merchandise and individuals were shipped, interfacing urban areas and districts, and working with the development of unrefined components and completed items.

Influence on Creation Effectiveness

The reception of iron hardware in plants significantly affected creation effectiveness. Motorization, fueled by steam motors and other iron-driven advancements, essentially sped up and result of assembling processes.

Iron hardware, because of its solidarity and strength, was instrumental in the formation of bigger and more productive machines. These machines could work at higher rates and for longer lengths, prompting significant additions underway proficiency. The automation of processing plants diminished the requirement for difficult work and considered the reliable and quick creation of products.

The division of work inside processing plants, made conceivable by the utilization of iron apparatus, further better productivity. Laborers became had some expertise specifically undertakings, like working a particular machine or dealing with a specific phase of the creation interaction. This specialization decreased the time expected to get done with every responsibility and added to a more smoothed out creation process.

Iron apparatus additionally worked with the formation of more solid and normalized creation strategies. The actual apparatus, built with accuracy parts, offered consistency and unwavering quality in the assembling system. Exchangeable parts, another development made conceivable by iron hardware, worked on the gathering and fix of machines, lessening margin time and expanding creation productivity.

The automation of creation processes and the utilization of iron hardware considered more noteworthy command over item quality. Normalized creation strategies, worked with by iron hardware, guaranteed that items satisfied reliable quality guidelines. This was especially vital in businesses like materials, where consistency was profoundly esteemed.

Expanded creation productivity, a sign of the plant framework, prompted higher result and lower creation costs. The capacity to create products for a bigger scope considered economies of scale, decreasing the expense per unit. This made items more reasonable as well as set out open doors for producers to grow their business sectors and arrive at a more extensive client base.

Influence on Work and Society

The reception of iron hardware in production lines had huge ramifications for work and society. The processing plant framework, made conceivable by iron-driven automation, reshaped work elements, prompted changes in working circumstances, and provoked the rise of work developments.

The progress from limited scope studios to production lines denoted a change in the idea of work. Assembly line laborers, who worked machines and performed specific undertakings, were important for a controlled timetable with fixed working hours. The production line framework presented another division of work, which was an obvious takeoff from the distinctive and more independent work of pre-modern settings.

The plant framework was described by concentrated creation offices that necessary a huge and restrained labor force. Laborers frequently had little independence and were supposed to stick to a relentless timetable. While motorization prompted higher efficiency, it likewise prompted a deskilling of specific undertakings, as assembly line laborers became had some expertise in restricted parts of creation.

The industrial facility framework's ascent additionally achieved new difficulties for laborers. Long working hours, frequently surpassing 12 hours out of every day, were normal in processing plants. The dull and tedious nature of manufacturing plant work, joined with swarmed and in some cases risky working circumstances, negatively affected the physical and mental prosperity of workers.

While the plant framework achieved more steady and unsurprising wages, laborers frequently confronted testing conditions. Production line proprietors looked to amplify benefits by limiting work costs, which frequently brought about low wages and unfortunate working circumstances. Work debates and strikes were normal as laborers squeezed for better wages, more limited working hours, and further developed security guidelines.

The development of work developments and associations was an immediate reaction to the difficulties looked by assembly line laborers. Laborers coordinated to advocate for their privileges and work on their functioning circumstances. Worker's guilds and laborers' freedoms associations battled for better wages, more limited business days, and more secure working circumstances, with a definitive objective of tending to the inconsistencies and difficulties experienced by workers.

The shift to the processing plant framework likewise impacted examples of movement. As provincial populaces moved to metropolitan focuses looking for work in the developing processing plants, urban areas extended quickly. Urbanization turned into a characterizing element of the Modern Upset, and it carried new provokes and potential open doors to the metropolitan scene. The development of modern urban areas changed the elements of lodging, public administrations, and social designs.

Influence on Urbanization and Framework

The reconciliation of iron hardware in production lines changed assembling processes as well as assumed a critical part in the extension of urbanization and the improvement of framework. The development of processing plants, controlled by steam motors and outfitted with iron hardware, was instrumental in reshaping the metropolitan scene.

Industrial facilities required enormous, meticulously designed designs to house apparatus and work. These structures, frequently developed with iron pillars and segments, included enormous open spaces to oblige apparatus and creation processes. The plan of modern structures took into consideration proficient formats, smoothing out creation and supporting bigger labor forces.

Modern urban communities arose as production lines multiplied. Urbanization turned into a focal element of the Modern Upheaval, with the development of urban areas driven by the interest for work in the plants. Laborers moved from rustic regions to metropolitan focuses looking for business, and this convergence of individuals essentially changed the segment structure of urban areas.

The development of modern urban communities likewise prompted the extension of framework and public administrations. Metropolitan focuses expected satisfactory transportation frameworks to help the development of individuals and merchandise. The improvement of rail organizations, made conceivable by steam trains and iron rails, associated urban communities and locales, working with the transportation of unrefined substances, completed items, and populaces.

Chapter 3

Architectural Marvels

Engineering, the workmanship and study of planning and developing structures, has been a significant articulation of human inventiveness and imagination since the beginning of time. From the amazing sanctuaries of old human advancements to the taking off high rises of the cutting edge period, design reflects our requirement for cover as well as our goals, convictions, and social personalities. This article celebrates compositional wonders from different ages, inspecting the verifiable and social settings that led to them and the getting through influence they have had on our reality.

The Pyramids of Giza: A Demonstration of Old Designing

The Pyramids of Giza, situated on the Giza Level in Egypt, are among the most notorious structural wonders ever. These titanic designs were developed during the Old Realm time frame, close to quite a while back. The Giza Pyramids comprise of three principal pyramids: the Incomparable Pyramid of Khufu (Cheops), the Pyramid of Khafre, and the Pyramid of Menkaure.

The development of the Incomparable Pyramid of Khufu, the biggest and generally renowned of the three, is a demonstration of old Egyptian designing and engineering ability. It is assessed that the Incomparable Pyramid comprises of around 2.3 million limestone and stone blocks, each gauging a few tons. These blocks were definitely quarried, shipped, and fitted along with amazing accuracy.

Hypotheses about the development methods of the pyramids proliferate, with some recommending the utilization of slopes, while others propose a more intricate framework including stabilizers and switches. No matter what the techniques utilized, the accomplishment of building such fantastic designs without current hardware is a demonstration of the expertise, association, and assurance of the old Egyptians.

The Giza Pyramids filled in as burial chambers for pharaohs and were intended to work with the progress of the departed to eternity. The Incomparable Pyramid, initially clad in cleaned Tura limestone packaging, mirrored the beams of the sun and was accepted to associate the natural domain with the heavenly. The accuracy of the pyramids' arrangements with divine bodies, like the North Star, features the high level galactic information on the old Egyptians.

The Parthenon: A Reference point of Old style Greek Engineering

The Parthenon, a sanctuary committed to the goddess Athena, remains as an image of old style Greek engineering and is arranged on the Acropolis in Athens, Greece. It was developed somewhere in the range of 447 and 438 BC during the Athenian Brilliant Age and was planned by the famous planner Ictinus, with the oversight of artist Phidias.

The Parthenon's compositional greatness lies in its Doric request, portrayed by fluted sections, entablatures, and pediments. The structure's extents stick to the standards of traditional Greek engineering, utilizing the Brilliant Proportion and the utilization of optical deceptions to address visual twists. The unobtrusive shape of the sanctuary's segments and its base checks the deception of hanging and guarantee a feeling of flawlessness.

The models that decorated the Parthenon were made by Phidias and were considered among the best instances of old style Greek workmanship. The pediments portrayed legendary scenes, while the metopes showed chivalrous fights, most quite the Clash of Long distance race and the Trojan Conflict. The Parthenon was a feature of the Greeks' creative and building ability, serving both as a strict safe-haven and an image of Athenian social and political accomplishments.

Throughout the long term, the Parthenon went through changes, from a sanctuary to a Christian church and, later, an Ottoman mosque. Its endurance through different times of history and its careful remaking endeavors in the twentieth century feature its getting through significance and the worldwide acknowledgment of its design importance.

The Colosseum: A Victory of Roman Designing

The Colosseum, otherwise called the Flavian Amphitheater, remains as a getting through image of Roman designing and engineering. Situated in the core of Rome, Italy, the Colosseum was built somewhere in the range of 70 and 80 Promotion during the rule of Sovereign Vespasian, with additional adjustments made during the standard of his child, Titus.

This huge amphitheater could hold an expected 50,000 to 80,000 onlookers and was utilized for different occasions, including gladiatorial challenges, creature chases, and dramatic exhibitions. The Colosseum's curved shape, with its arrangement of layered seating and underground passages, took into consideration proficient group the board and guaranteed fast section and exit.

The Colosseum's most unmistakable element is its utilization of curves and vaults, a compositional advancement that added to its security and glory. The utilization of cement and travertine stones considered the development of such monstrous designs. The field's wooden floor, covered with sand, could be taken out to uncover the twisted passages underneath, lodging fighters, creatures, and stage sets.

The Colosseum's plan exemplified Roman designing standards, displaying the authority of curves and vaults. The broad arrangement of passages, hidden entryways, and lifts gave a way to intricate and emotional doorways and ways out for the two entertainers and creatures. The Colosseum's proficient plan and limit with regards to enormous crowds were a demonstration of the Romans' designing discernment.

Notwithstanding experiencing harm seismic tremors, fires, and the looting of its stone for different developments, the Colosseum has persevered as a structural wonder, drawing in great many guests yearly and filling in as a demonstration of the loftiness and refinement of Roman engineering.

The Incomparable Mass of China: A Landmark to Safeguard and Solidarity

The Incomparable Mass of China, one of the most famous design miracles of the world, addresses a remarkable accomplishment of old designing and development. Extending more than 13,000 miles across northern China, it's anything but a solitary nonstop wall however a progression of fortresses worked by various traditions over hundreds of years.

The starting points of the Incomparable Wall date back to the seventh century BC, during the Fighting States time frame in China. At first, individual states built walls for safeguard against trespassers. It was during the Qin Tradition (221-206 BC) that Sovereign Qin Shi Huang started the unification of these walls into a solitary framework, denoting the start of the Incomparable Wall.

The Incomparable Wall's essential capability was guarded, safeguarding China from the different roaming clans and trespassers from the north. The wall, developed principally with earth, wood, and stone, was strengthened with lookouts, signal pinnacles, and post stations. The utilization of packed earth, stone, and blocks took into consideration a strong and versatile construction equipped for enduring everyday hardship.

The scale and intricacy of the Incomparable Wall are faltering, as it navigates various landscapes, including deserts, mountains, and meadows. The wall's essential areas, lookouts, and sign flames considered quick correspondence and reaction to expected dangers.

The Incomparable Wall likewise filled in for of controlling exchange and migration along the Silk Street and defending the progression of merchandise and data. The Ming Line (1368-1644) further extended and built up the Incomparable Wall, adding to its ongoing notable appearance.

3.1 Iron in Architecture Throughout History

Iron, an essential metal of striking strength and adaptability, plays had a necessary impact in structural history for a really long time. From the earliest ironwork in antiquated designs to the famous iron milestones of the modern age, this paper investigates the advancing utilization of iron in engineering and the persevering through effect of this material on the assembled climate.

Early Use of Iron in Engineering

The historical backdrop of iron in design can be followed back to old civilizations where iron was utilized in different structures. One of the earliest purposes of iron in engineering is found in the Hittite human advancement of Anatolia (current Turkey) around 1300 BC. The Hittites utilized iron stakes to join stone blocks in development, denoting a huge early use of iron as a structure material. Iron's sturdiness and strength made it an alluring choice for making associations that could endure everyday hardship.

The Romans, eminent for their structural and designing accomplishments, likewise embraced the utilization of iron in development. The approach of iron clips and dowels permitted the Romans to support their enormous stone designs, including water passages, scaffolds, and sanctuaries. Iron was utilized as both support and connectors, guaranteeing the solidness and life span of these design ponders. The Pantheon in Rome, worked around 126 Promotion, remains as a getting through illustration of Roman design that consolidates iron clasps in its substantial arch construction, adding to its striking protection throughout the long term.

Iron in Middle age and Gothic Engineering

As Europe progressed into the middle age period, iron kept on assuming a crucial part in engineering. One remarkable application was the utilization of iron tie bars in the development of Gothic houses of God. These churches, with their transcending towers and far reaching windows, presented special underlying difficulties. The outward pushed of the ribbed vaults and the huge load of stone required inventive answers for solidness.

Iron tie poles were utilized to neutralize the sidelong powers and keep the walls from spreading separated. These tie poles went about as pressure individuals, arranging the walls and taking into consideration the taking off, open insides normal for Gothic engineering. One of the most popular models is the Notre-Lady House of prayer in Paris, which consolidates iron tie poles in its primary framework.

Notwithstanding their primary capability, iron tie bars in Gothic houses of prayer were frequently coordinated into the general stylish. They were here and there left uncovered and embellished with improving components, filling in as both practical backings and fancy highlights.

The Renaissance and Iron's Ornamental Job

The Renaissance time frame denoted a re-visitation of traditional design standards, drawing motivation from the engineering of old Greece and Rome. During this time, iron kept on being utilized primarily, however it additionally started to fill enlivening needs.

One remarkable improvement was the utilization of iron galleries and railings in Renaissance design. These elaborate ironwork components were frequently incorporated into the façades of structures, adding a feeling of style and elegance. The complicated plans of these overhangs and railings exhibited the creativity of ironwork, with examples and themes that went from flower themes to mathematical shapes. Ironwork turned into a method for decorating structural veneers, especially in districts like Italy and France.

One of the most well known instances of enhancing ironwork in Renaissance design is the Juliet Overhang in Verona, Italy, related with the amazing gallery from Shakespeare's "Romeo and Juliet." The fancy ironwork of this gallery, decorated with smooth bends and flower designs, epitomizes the combination of craftsmanship and engineering in this period.

Iron in the Modern Age: An Upset in Design

The nineteenth century achieved an emotional change in design, driven by the modern upset and the boundless utilization of iron. The advancement of new iron creation methods, for example, the Bessemer cycle, made iron all the more promptly accessible and reasonable. Subsequently, iron turned into a central material in structural development.

One of the most notorious building wonders of the modern age is the Precious stone Castle, planned by Sir Joseph Paxton for the Incomparable Presentation of 1851 in London. The Gem Castle was an extreme takeoff from conventional engineering styles, highlighting a tremendous glass and iron construction that exemplified the capability of these materials. Iron assumed a focal part in the royal residence's development, filling in as the system for the far reaching glass boards that wrapped the structure. This imaginative utilization of iron and glass exhibited the potential outcomes of these materials in making huge, open, and sufficiently bright spaces.

The Eiffel Pinnacle, finished in 1889 for the Article Universelle in Paris, remains as one more notorious illustration of iron in engineering. Planned by Gustave Eiffel, the pinnacle's many-sided iron grid structure was a demonstration of the designing capacities of the time. The pinnacle's iron structure, which ascends to a level of 324 meters, was an exceptional accomplishment of development. The Eiffel Pinnacle's plan featured the tasteful capability of iron, showing the way that this utilitarian material could likewise be utilized to make outwardly striking and getting through compositional milestones.

Iron's job in design was not restricted to terrific displays and milestones. The material found boundless application in the development of modern offices, train stations, and extensions. Iron scaffolds, with their complicated cross section plans and iron supports, turned into a sign of nineteenth century designing. The iron rail line stations of this time additionally highlighted lavish ironwork, displaying the combination of structure and capability in design.

The Advancement of Iron in Current Engineering

The utilization of iron in design kept on advancing in the twentieth hundred years. The pioneer development embraced the stylish capability of modern materials, including iron and steel. Engineers, for example, Ludwig Mies van der Rohe and Le Corbusier integrated steel and iron into their plans, making structures described by clean lines, moderate structures, and an accentuation on functionalism.

The Seagram Building, planned by Mies van der Rohe and finished in 1958 in New York City, is a commendable portrayal of this methodology. The structure's smooth steel and glass façade, with its moderate plan, exemplifies the pioneer vision of straightforwardness and effortlessness in design. The utilization of steel and iron in the primary system considered the formation of enormous, segment free inside spaces, adding to the structure's useful style.

The Sydney Drama House, planned by Jørn Utzon and finished in 1973, is one more milestone that features the flexibility of iron in present day engineering. The structure's unmistakable sail-like shells are developed utilizing precast substantial boards built up with iron bars. The creative plan of the Drama House, joined with its famous appearance, has made it an image of Australia and an UNESCO World Legacy Site.

In ongoing many years, iron and steel have kept on being essential to compositional development. Contemporary planners have investigated new structures, materials, and innovations that influence the strength and adaptability of these metals. From the imaginative utilization of enduring steel in the plan of structures to the consolidation of iron in supportable development, the advancement of iron in engineering keeps on forming the assembled climate.

Supportable Practices and Iron in Design

The 21st century has gotten a developing spotlight on supportability engineering, and iron assumes a part in this shift. Maintainability in design envelops different perspectives, including energy productivity, material reuse, and ecological effect. Iron, as a profoundly recyclable material, lines up with these manageability objectives.

Reused iron, frequently as reused steel, is progressively utilized in development. The reuse of iron and steel saves normal assets as well as decreases energy utilization and ozone depleting substance emanations related with iron creation. This maintainable practice is especially applicable in the development of current structures and framework, where natural contemplations are fundamental.

The feasible utilization of iron in engineering reaches out past material decision. Iron can likewise be coordinated into the plan of harmless to the ecosystem structures. For instance, iron screens or louvers can be utilized to control regular light and ventilation, decreasing the energy expected for warming, cooling, and fake lighting. Also, iron and steel outlining frameworks can be intended to advance energy proficiency and warm execution.

Difficulties and Contemplations in Involving Iron in Engineering

While iron offers various advantages in design, including its solidarity, recyclability, and tasteful flexibility, there are likewise difficulties to consider. Iron can erode over the long haul when presented to the components, prompting upkeep and reclamation prerequisites. Appropriate coatings and completions are many times important to shield iron components from rust and rot.

The heaviness of iron can likewise be a thought in compositional plan, particularly in tall or huge scope structures. Specialists and designers should cautiously ascertain and anticipate the heap bearing limit of iron parts. At times, the utilization of steel, a lighter and more grounded combination of iron, might be ideal for underlying components.

Moreover, the expense of iron and steel, while serious with other structure materials, can vary in light of economic situations and worldwide stock chains. Engineers and developers should represent these variables in project spending plans and timetables.

3.2 The Eiffel Tower: A Symbol of Iron's Aesthetic Potential

The Eiffel Pinnacle, perhaps of the most famous milestone on the planet, isn't simply a transcending design of iron and steel; it is a demonstration of the stylish capability of these materials. Transcending the city of Paris, the Eiffel Pinnacle has caught the creative mind of millions since its development for the 1889 Article Universelle (World's Fair). This exposition investigates the set of experiences, plan, and social meaning of the Eiffel Pinnacle, featuring its job as an image of iron's stylish excellence and designing wonder.

A Victory of Plan and Designing

Gustave Eiffel, a French structural designer, and his underlying specialists, Maurice Koechlin and Émile Nouguier, planned the Eiffel Pinnacle as the highlight of the 1889 Work Universelle, held in Paris to praise the 100th commemoration of the French Upheaval. The pinnacle was expected to grandstand French modern ability and designing greatness, and it did as such with unrivaled effortlessness.

The plan of the Eiffel Pinnacle is described by its open iron cross section structure, which is both mind boggling and useful. Remaining at a level of 324 meters (1,063 feet) when it was finished, it was the tallest man-made structure on the planet at that point. The pinnacle's level, accomplished through its iron grid configuration, permitted it to act as a noticeable point of convergence for the work.

The pinnacle's design comprises of in excess of 18,000 individual iron parts, associated by over 2.5 million bolts. These parts incorporate iron bars and latticed curves that structure the pinnacle's notorious outline. The utilization of iron considered the production of a lightweight yet uncommonly solid design, a sign of Eiffel's designing development.

The Eiffel Pinnacle's iron construction is partitioned into three particular levels. The main level, open to guests by means of steps or a lift, contains various cafés, shops, and review regions. The subsequent level, arranged at a level of 115 meters (377 feet), offers stunning all encompassing perspectives on Paris. The third level, arriving at a level of 276 meters (906 feet), gives a much more raised point of view, complete with a glass floor for a genuinely vertiginous encounter.

The development of the Eiffel Pinnacle introduced various designing difficulties, including the exact arrangement and get together of its iron parts. To accomplish this, Eiffel and his group fostered a clever arrangement of construction, where iron components were produced off-site, set apart with remarkable recognizable proof numbers, and afterward shipped to the building site for gathering. This technique guaranteed accuracy and effectiveness during the development interaction.

The gathering of the pinnacle's iron parts was executed with wonderful precision. The utilization of bolts, a typical securing technique in iron and steel development, considered secure and sturdy associations between the components. The cautious execution of the iron grid structure brought about a pinnacle that remained steadfast and valid, even after over 100 years of presence.

A Getting through Image of Paris

The Eiffel Pinnacle was finished in 1889 and was met with both praise and analysis. While certain Parisians at first saw it as a dubious and eccentric expansion to the city's horizon, it before long turned into a persevering through image of Paris and France itself. The pinnacle's rich structure and creative plan caught the creative mind of people in general and raised the city's status as a focal point of craftsmanship, culture, and designing.

The iron grid structure, which at first drew suspicion, was praised as a spearheading compositional accomplishment. It displayed the capability of iron as a development material, testing assumptions of what was conceivable in building plan. The Eiffel Pinnacle's elegant structure, which tightens as it climbs, conveys a feeling of daintiness and smoothness, challenging its monstrous iron development.

One of the pinnacle's most convincing highlights is the utilization of iron to make a mind boggling play of light and shadow. The cross section structure, especially when enlightened around evening time, changes the pinnacle into an enrapturing reference point that intersperses the Parisian horizon. The pinnacle's evening light upgrades its stylish allure and adds a feeling of sentiment to the city of Paris.

The Eiffel Pinnacle likewise assumed a critical part during its initial a very long time in the improvement of broadcast communications. In 1903, the pinnacle's highest point was furnished with a remote telecommunication station, denoting a mechanical achievement throughout the entire existence of correspondence. This utilization of the pinnacle featured its flexibility and versatility, displaying iron's possible past its tasteful characteristics.

Social Importance and Inheritance

The Eiffel Pinnacle's social importance stretches out past its designing and engineering achievements. It has turned into a persevering through image of French culture and a portrayal of Paris itself. The pinnacle's persevering through fame has made it a worldwide symbol and a strong epitome of the City of Light.

Over time, the Eiffel Pinnacle has been highlighted in endless movies, writing, and masterpieces. It has become inseparable from sentiment, bringing out pictures of darlings sharing a kiss under its curves. The pinnacle's social reverberation has made it an image of Parisian class and refinement.

The pinnacle has likewise assumed a critical part in worldwide occasions and memorable minutes. During both Universal Conflicts, the Eiffel Pinnacle filled in as an essential radio transmission point and was basic to correspondences during those wild times. It stays an image of strength and perseverance, having endured different difficulties all through its presence.

In the domain of craftsmanship and engineering, the Eiffel Pinnacle has affected endless works and specialists. Its grid structure, which consolidations structure and capability, has propelled modelers and fashioners looking to push the limits of what is conceivable in compositional style. The pinnacle's impact should be visible in the grid and steel plans of contemporary structures around the world.

Continuous Conservation and Transformation

Saving the Eiffel Pinnacle has been a continuous exertion since its finishing. The iron parts, presented to the components, have been helpless to consumption and enduring. Customary support and reclamation work are fundamental to guaranteeing the pinnacle's underlying uprightness and tasteful allure.

To shield the pinnacle from consumption, it has gone through various paint applications throughout the long term. The notable ruddy earthy colored shade of the pinnacle is a consequence of its utilization of "Eiffel Pinnacle Brown" paint. A group of painters intermittently applies another layer of paint, which requires a while to finish and uses an expected 60 metric lots of paint.

The Eiffel Pinnacle has additionally seen different variations and developments to improve its usefulness and availability. Lifts and other specialized establishments have been refreshed and modernized to oblige the requirements of contemporary guests. Moreover, the pinnacle has presented economical practices, for example, energy-productive lighting and waste decrease endeavors, adjusting it to current natural and preservation objectives.

One outstanding variation is the establishment of straightforward glass floors on the pinnacle's most memorable level, giving guests a one of a kind and thrilling experience. The glass floors offer unhindered perspectives on the ground underneath, making a feeling of drifting over the city. This variation is a demonstration of the pinnacle's capacity to join its verifiable importance with current developments, showing its proceeded with significance in the 21st 100 years.

The Eiffel Pinnacle in the 21st Hundred years

The Eiffel Pinnacle stays a social and structural wonder, drawing in large number of guests from around the world every year. It remains as a getting through image of Paris, France, and human creativity. In the 21st hundred years, the pinnacle proceeds to motivate and charm, filling in as an impression of the getting through allure of iron in building plan.

The Eiffel Pinnacle's inheritance reaches out past its visual magnificence and designing greatness. It encapsulates the soul of advancement and the limit of compositional plan to rise above its utilitarian reason. The pinnacle's iron grid structure, which once tested show, has come to represent the agreeable marriage of structure and capability.

As a worldwide symbol, the Eiffel Pinnacle stays a wellspring of motivation for modelers, specialists, and craftsmen. Its complex ironwork, enlightened presence, and amazing vistas proceed to captivate and elevate the people who experience it. The pinnacle's ability to join custom and innovation, strength and elegance, addresses the persevering through allure of iron as a material of boundless likely in the realm of engineering.

During a time of quick mechanical progression and consistently developing building patterns, the Eiffel Pinnacle remains as an immortal indication of the enduring effect that inventive plan and designing can have on the fabricated climate. It is a recognition for the vision of Gustave Eiffel and the incalculable people who added to its creation, and it stays an image of the unfathomable conceivable outcomes that iron and steel proposition to engineering plan and articulation.

3.3 Modern Architectural Innovations

Present day engineering is a powerful field portrayed by imaginative thoughts, state of the art innovations, and a pledge to pushing the limits of plan. In the 21st hundred years, planners and creators are continually looking for better approaches to address the developing requirements of society, make practical and energy-productive designs, and embrace arising advancements. This paper digs into different present day design developments that have changed the manner in which we contemplate the fabricated climate.

Supportability and Green Design

Supportability is a focal subject in current engineering, driven by the pressing need to address natural worries, for example, environmental change and asset exhaustion. Green engineering includes a scope of imaginative practices and innovations pointed toward decreasing a structure's natural effect and further developing its energy proficiency.

One noticeable advancement is the incorporation of sustainable power frameworks into structures. Sunlight powered chargers, wind turbines, and geothermal intensity siphons are currently generally used to produce clean energy on location. These frameworks decrease a structure's carbon impression as well as add to energy freedom and long haul cost investment funds.

Moreover, engineers are embracing inactive plan standards to improve a structure's energy proficiency. Procedures like upgrading regular light, utilizing latent sun powered warming and cooling, and consolidating normal ventilation can fundamentally lessen a design's dependence on mechanical frameworks and lower its energy utilization.

Materials determination assumes a pivotal part in reasonable design. Advancements in material science have prompted the improvement of eco-accommodating structure materials, for example, low-influence concrete, reused steel, and maintainable wood items. These materials offer superior strength and diminished ecological effect contrasted with conventional other options.

The appearance of green rooftops and living walls is one more creative practice in present day engineering. These highlights upgrade a structure's energy effectiveness, decrease stormwater spillover, and make metropolitan desert gardens that help biodiversity. Green rooftops, specifically, give protection and assist with relieving the metropolitan intensity island impact.

Versatile Reuse and Notable Protection

Current modelers are progressively perceiving the worth of versatile reuse and memorable conservation in establishing manageable and dynamic metropolitan conditions. Versatile reuse includes reusing existing designs for new capabilities, reinvigorating noteworthy structures, and lessening the interest for new development.

Imaginative versatile reuse projects incorporate changing over modern stockrooms into private lofts, changing notable schools into public venues, and reusing old industrial facilities into inventive office spaces. These activities save the person and social legacy of existing structures while limiting the ecological effect of destruction and new development.

Current design rehearses frequently include consistently coordinating old and new components inside a venture. This approach consolidates the noteworthy beguile of a structure with contemporary plan components and innovations, making novel and dynamic spaces.

Parametric Plan and Advanced Manufacture

Parametric plan is a state of the art building approach that use computational calculations to produce mind boggling, variable, and profoundly redid plans. Designers utilize parametric programming to make unpredictable, natural structures that were once challenging to imagine and build. Parametric plan empowers draftsmen to streamline structures for explicit capabilities, like expanding regular light or acoustics, and to make stylishly staggering structures.

Advanced manufacture strategies, including 3D printing, CNC (PC Mathematical Control) processing, and automated gathering, have reformed the manner in which planners rejuvenate their plans. These advancements take into account exact, proficient, and financially savvy assembling of complicated building parts.

One remarkable illustration of parametric plan and computerized manufacture is the Serpentine Structure in London, planned by designers like Zaha Hadid and Bjarke Ingels. These structures feature the capability of advanced plan and manufacture to make liquid, sculptural designs that oppose conventional development strategies.

Biophilic Plan and Wellbeing Engineering

Biophilic configuration is an inventive methodology that coordinates nature into the constructed climate, making spaces that advance prosperity and network with the regular world. Modelers use components like normal materials, vegetation, and sunshine to improve the physical and mental wellbeing of building inhabitants.

The WELL Structure Standard, which centers around establishing sound indoor conditions, has built up forward momentum lately. This standard envelops plan components that upgrade air and water quality, lighting, wellness, and mental prosperity, bringing about spaces that focus on the wellbeing and solace of their clients.

The effect of biophilic plan and wellbeing engineering stretches out to different structure types, from workplaces and medical care offices to private homes. It mirrors a developing consciousness of the significant association between our environmental elements and our physical and emotional wellness.

Particular and Pre-assembled Development

Particular and pre-assembled development techniques are reshaping how structures are planned and built. These methodologies include the gathering of building parts off-site in a controlled climate, offering a few benefits, including diminished development time, cost reserve funds, and worked on quality control.

One outstanding illustration of secluded development is the rising prominence of measured lodging units. These units, worked in production lines, can be stacked, consolidated, and tweaked to make private or business structures. They offer answers for lodging deficiencies, assist development timetables, and diminish squander.

Super advanced and Shrewd Structures

The ascent of super advanced and shrewd structures is one of the main design developments of the cutting edge period. These designs integrate cutting edge innovations that upgrade their usefulness, productivity, and supportability. Brilliant structures use sensors, mechanization, and information investigation to advance energy use, security, and client solace.

With regards to cutting edge engineering, the famous Apple Park grounds in Cupertino, California, planned by Cultivate + Accomplices, is a surprising model. The grounds includes a round plan with a huge, bended glass façade, making a consistent association between the structure and the encompassing scene.

The design consolidates energy-proficient air conditioning frameworks, normal ventilation, and on location environmentally friendly power sources, making it a model of practical, innovative engineering.

Drifting and Submerged Engineering

As land becomes more difficult to find in metropolitan regions, modelers are investigating imaginative answers for grow our constructed climate. Drifting and submerged design addresses an imaginative reaction to the difficulties of land shortage and rising ocean levels.

Drifting structures, like houseboats and drifting inns, are intended to rise and fall with changing water levels. These designs give special waterfront living encounters while exhibiting the flexibility of engineering in an evolving environment.

Submerged design, however still in its earliest stages, holds the commitment of making lowered natural surroundings for different purposes, including exploration, the travel industry, and hydroponics. Trial projects like the Water Disk Lodging in Dubai and the Sea Winding idea by Shimizu Enterprise in Japan allude to a future where structures might flourish underneath the sea's surface.

Vertical Backwoods and Skybridges

Metropolitan conditions frequently need green spaces, which is a worry for the two feel and ecological wellbeing. Vertical backwoods, as found in projects like the Bosco Verticale in Milan by Stefano Boeri, include establishing trees and vegetation on the façades of tall structures. These "living high rises" give vegetation, lessen CO2 outflows, and further develop air quality in thickly populated urban areas.

Skybridges, or flying walkways interfacing structures, are another building development. These raised entries work with walker development between structures as well as proposition special spaces for social cooperation and recreation. Notable models remember the High Line for New York City, which has changed an old rail track into a raised park, and the Marina Straight Sands SkyPark in Singapore, which flaunts a 150-meter-long vastness pool and all encompassing perspectives on the city.

Origami and Active Design

Origami, the old craft of paper collapsing, has roused another rush of compositional development. Origami-based plan standards are being utilized to make active designs that can change shape, extend, or move in light of natural circumstances or client needs.

Dynamic engineering incorporates components like retractable rooftops, collapsing façades, and expandable designs. These elements permit structures to adjust to evolving climate, advance energy use, and improve client encounters. For instance, arenas with retractable rooftops can switch between outdoors and encased arrangements in view of atmospheric conditions, while motor façades can conform to control light and ventilation.

Metropolitan Cultivating and Vertical Agribusiness

The developing interest in practical food creation has prompted advancements in metropolitan cultivating and vertical farming. These practices include developing yields, organic products, and vegetables inside the city, utilizing restricted space and creative methods like aquaculture and aeroponics.

Quite possibly of the most aggressive task in this class is the Upward Ranch in Linköping, Sweden. This 16-story vertical ranch, planned by Plantagon, utilizes robotized frameworks to develop vegetables and spices all year. Vertical agribusiness can lessen the carbon impression of food creation, further develop food security.

Chapter 4

Iron and Transportation

Iron, with its exceptional strength and flexibility, plays had a urgent impact in changing transportation frameworks since the beginning of time. From the earliest iron wheels to the improvement of iron scaffolds, rail routes, and vehicles, this paper investigates the getting through effect of iron on the development of transportation, producing a cutting edge universe of portability and network.

The Beginning of Iron in Transportation

The use of iron in transportation can be followed back to antiquated times, denoting the progress from additional crude materials to the utilization of metals for wheels, axles, and different parts. This shift changed the proficiency and solidness of early methods of transport.

One of the earliest known instances of iron's application in transportation is the utilization of iron-rimmed wheels on horse-attracted chariots antiquated China. The consolidation of iron edges essentially worked on the strength and execution of these vehicles, permitting them to navigate harsh landscapes and significant distances no sweat. This early utilization of iron established the groundwork for the improvement of further developed transportation frameworks in the hundreds of years to come.

Iron in Sea Transport

Iron likewise assumed a significant part in sea transport, especially in the development of boats and maritime vessels. The progress from wooden to press shipbuilding during the nineteenth century denoted a huge headway in transport plan and execution.

The presentation of iron and later steel bodies permitted boats to be bigger, more hearty, and less vulnerable to the mileage that impacted wooden vessels. The utilization of iron and steel achieved another period of shipbuilding, prompting the making of strong steamships that altered overseas travel and worldwide exchange.

The SS Extraordinary Eastern, planned by Isambard Realm Brunel and sent off in 1858, is a remarkable illustration of this shift to press in sea transport. With an iron body, the SS Incredible Eastern was the biggest boat of now is the ideal time, fit for conveying travelers and freight on lengthy journeys across the Atlantic. This pivotal vessel set up for the advancement of current sea liners and freight ships.

The Iron Pony: Rail routes and Trains

Maybe perhaps of the most extraordinary improvement in transportation history was the appearance of the rail route, fueled by steam trains. The utilization of iron, and later steel, in rail line tracks, extensions, and trains laid the preparation for the development of rail networks that associated urban areas and locales, altering both traveler and cargo transport.

George Stephenson, known as the "Father of Rail routes," planned the main fruitful steam train, the Rocket, in 1829. This iron-wheeled train set new guidelines for speed and proficiency in transportation. Iron rails were likewise presented, lessening contact and mileage on tracks. The blend of iron tracks and trains denoted a critical jump forward in the dependability and speed of transportation.

Iron scaffolds, like the Iron Extension in Shropshire, Britain, finished in 1779, exhibited the possibility of iron designs in rail line development. Iron was appropriate to the requests of railroad spans, offering the strength and sturdiness expected to help the weighty heaps of trains and moving stock.

As rail lines extended internationally, iron turned into a focal part of the transportation foundation. It worked with the development of individuals and products over huge distances, associating recently separated locales and adding to the monetary and social advancement of countries.

The Cross-country Railroad and the American West

In the US, the development of the Principal Cross-country Railroad during the 1860s denoted a noteworthy crossroads in transportation history. Iron and steel assumed a crucial part in the improvement of this fantastic venture, which connected the eastern and western shores of the nation and changed the American West.

The Association Pacific and Focal Pacific Rail lines, the two organizations answerable for building the cross-country course, utilized iron and steel rails, extensions, and passages to navigate the difficult territory of the western US. The utilization of iron and steel empowered the development of a persistent rail line that decisively decreased travel time and expenses, while likewise encouraging financial improvement along the course.

The Brilliant Spike, crashed into the ground at Projection Highest point, Utah, in 1869, denoted the finish of the Main Cross-country Railroad, representing the unification of the US and the extraordinary force of iron in transportation.

Crossing over Distances with Iron: The Coming of Iron Extensions

Iron extensions address one more critical utilization of iron in transportation. These designs play had an essential impact in associating networks, crossing waterways, and spreading over testing scenes.

The improvement of iron scaffold development was advanced in the nineteenth hundred years by propels in iron assembling procedures, for example, the Bessemer cycle, which made iron creation more productive and reasonable. This prompted the development of a huge number of iron extensions that added to the development and improvement of districts all over the planet.

One eminent model is the Iron Scaffold in Shropshire, Britain, planned by Thomas Telford and finished in 1779. This extension, with its exquisite curves and unpredictable ironwork, filled in as an early exhibition of iron's likely in structural designing. The Iron Scaffold was a useful intersection as well as an image of the modern age and the versatility of iron in development.

Iron bracket spans, portrayed by their cross section like plan, turned into a sign of nineteenth century designing. These extensions highlighted unpredictable ironwork that gave both strength and feel. The Brooklyn Extension, planned by John A. Roebling and finished in 1883, is a noticeable illustration of this designing wonder. The extension's blend of steel links and iron trusswork made a strong and outwardly striking design that associated Manhattan and Brooklyn.

Iron scaffolds have not exclusively been fundamental for transportation yet have additionally added to the social and design scenes of urban communities and areas. Their getting through presence and useful excellence exhibit the significant effect of iron on the fabricated climate.

From Iron to Steel: The Change in Development Materials

While iron assumed a vital part in early transportation foundation, the change to steel denoted a huge headway. Steel, an amalgam of iron, offered more prominent strength and sturdiness, settling on it an optimal decision for building scaffolds, high rises, and other transportation-related structures.

The late nineteenth and mid twentieth hundreds of years saw the ascent of steel spans, which were praised for their proficiency, strength, and plan flexibility. The reception of steel permitted architects to make longer ranges, expanding the adaptability and proficiency of transportation organizations.

The Forward Scaffold in Scotland, finished in 1890, is a notorious steel span that exhibits the capability of this material. The scaffold's three cantilevered ranges, each with a focal range of 520 meters (1,710 feet), were developed utilizing 54,000 tons of steel. This designing accomplishment featured the capacities of steel in making huge, sturdy, and outwardly amazing designs.

The advancement of steel-supported concrete, an imaginative blend of steel and concrete, further extended the conceivable outcomes in transportation framework. This composite material offers both the strength of steel and the flexibility of cement, settling on it an optimal decision for developing extensions, passages, and interstates.

The Auto Upheaval: Iron and the Advanced Vehicle

Iron remaining parts a fundamental material in the auto business. The improvement of the vehicle in the late nineteenth and mid twentieth hundreds of years denoted a huge change in transportation, and iron assumed a key part in the development of auto plan and assembling.

The body and suspension of early cars were developed utilizing iron and steel parts. These materials offered the essential strength and inflexibility to endure the requests of street travel. The coming of sequential construction system creation, outstandingly spearheaded by Henry Passage, sped up the large scale manufacturing of vehicles and made them open to a more extensive populace.

As car innovation kept on propelling, iron and steel parts became lighter and more solid, adding to the improvement of more secure and more productive vehicles. Advancements in metallurgy and assembling methods have prompted the production of high-strength prepares that further develop vehicle crash execution and eco-friendliness.

Present day cars are furnished with different wellbeing highlights, including air-bags, safety belt pretensioners, and fold zones, a significant number of which are produced using iron and steel. These highlights have essentially improved traveler security and diminished the seriousness of auto crashes.

Fast Rail and Maglev Innovation

In ongoing many years, transportation has seen the rise of high velocity rail organizations and attractive levitation (maglev) innovation. These advancements in rail travel epitomize the continuous job of iron and steel in pushing the limits of speed, effectiveness, and manageability.

High velocity rail frameworks, which use committed tracks and smoothed out trains, have changed really long travel, offering an all the more harmless to the ecosystem choice to air travel and individual vehicles. The Shinkansen, otherwise called the "slug train," in Japan was one of the principal rapid rail frameworks, highlighting smooth and streamlined trains that movement at paces of as much as 320 kilometers each hour (200 miles each hour).

Maglev innovation takes rail travel to a higher level by killing the requirement for actual contact between the train and the track. Maglev trains, moved by attractive fields, can accomplish .

4.1 Iron's Role in Revolutionizing Transportation

The narrative of iron in transportation is a story of change, development, and progress. Iron, with its special mix of solidarity, strength, and flexibility, plays had an essential impact in the development of transportation frameworks since the beginning of time. From the early utilizations of iron in horse-attracted vehicles to the improvement of iron scaffolds, rail lines, cars, and fast trains, this exposition investigates how iron has been a main thrust behind the transformation of the world's transportation organizations.

Early Starting points: Iron on Haggles

The excursion of iron in transportation starts in old times, when iron was first acquainted with upgrade the presentation and toughness of wheeled vehicles. As a substitution for prior materials, iron offered a critical headway in the strength and versatility of transportation frameworks.

One of the earliest examples of iron in transportation is the utilization of iron-rimmed wheels on horse-attracted chariots old China. This brilliant use of iron gave chariots improved solidness, empowering them to explore rough territories and cover longer distances. Iron-rimmed wheels denoted a defining moment in early transportation, preparing for future developments.

Iron likewise assumed a significant part in the improvement of horseshoes, an innovation that superior the versatility and life span of ponies utilized for transportation. Iron horseshoes safeguarded the hooves of these creatures, diminishing wear and injury, and empowering them to travel more noteworthy distances. This headway had significant ramifications for land-based transportation, as it expanded the reach and effectiveness of pony drawn vehicles.

The Period of Investigation: Iron in Oceanic Vehicle

Oceanic vehicle was another field where iron tracked down essential applications. The change from wooden shipbuilding to press transport development during the nineteenth century denoted a stupendous change in sea innovation. Iron and later steel frames upset transport plan and execution, adding to the ascent of strong steamships that changed maritime travel and worldwide exchange.

One of the most praised instances of iron in oceanic vehicle is the SS Extraordinary Eastern, planned by the spearheading engineer Isambard Realm Brunel and sent off in 1858. This iron-hulled vessel was the biggest boat of now is the right time, fit for conveying travelers and freight on overseas journeys. The SS Extraordinary Eastern showed the tremendous capability of iron in shipbuilding, making way for the advancement of current sea liners and freight ships.

The change to press and steel in transport development gave ships a few benefits, including expanded size, improved heartiness, and more noteworthy protection from the mileage that impacted wooden vessels. This change denoted another period in shipbuilding, encouraging the production of steamships that were more proficient, solid, and appropriate for really long travel.

The Introduction of Railroads: Iron Tracks and Steam Trains

Quite possibly of the most extraordinary improvement in transportation history was the approach of the rail route, fueled by steam trains. The mix of iron, and later steel, into rail line tracks, scaffolds, and trains was instrumental in the extension of rail networks that associated urban areas, districts, and countries.

The improvement of the steam train addressed a jump forward in transportation proficiency. In 1829, George Stephenson, frequently alluded to as the "Father of Rail lines," planned the Rocket, the primary effective steam train. This iron-wheeled train set new norms for speed and effectiveness, changing how individuals and products were moved.

The reception of iron rails likewise assumed a critical part in the rail route unrest. Iron rails were less helpless to mileage than wooden tracks, diminishing rubbing and expanding the sturdiness of the rail route foundation. These iron tracks established the groundwork for the making of dependable and effective railroad frameworks.

Additionally, the development of iron extensions, like the notorious Iron Scaffold in Shropshire, Britain, finished in 1779, showed the possibility of involving iron designs in rail route development. Iron scaffolds were instrumental in offering the vital help for the weighty heaps of trains and moving stock, working with the advancement of broad railroad organizations.

As rail lines extended internationally, iron turned into a foundation of the transportation framework. It associated recently separated districts, worked with the development of individuals and products over significant distances, and added to the monetary and social advancement of countries.

The Cross-country Railroad and the American West

The US saw an extraordinary crossroads in transportation history with the development of the Primary Cross-country Railroad during the 1860s. This notable task, associating the eastern and western shorelines of the nation, assumed a vital part in the improvement of the American West.

The Association Pacific and Focal Pacific Railways, answerable for building the cross-country course, depended on iron and steel rails, scaffolds, and passages to explore the difficult landscape of the western US. The utilization of iron and steel empowered the development of a ceaseless rail line, lessening travel time and expenses and cultivating monetary improvement along the course.

The fulfillment of the Principal Cross-country Railroad was represented by the driving of the Brilliant Spike at Projection Highest point, Utah, in 1869. This notorious second denoted the unification of the US and featured the groundbreaking force of iron in transportation.

Spanning Holes: Iron Scaffolds as Engineering Wonders

Iron extensions address a critical utilization of iron in transportation and structural designing. These designs play had a crucial impact in interfacing networks, crossing waterways, and traversing testing scenes.

The eighteenth century Iron Scaffold in Shropshire, Britain, planned by Thomas Telford and finished in 1779, remains as an early illustration of iron's expected in structural designing. With its rich curves and multifaceted ironwork, the Iron Extension was a useful intersection as well as an image of the modern age and the flexibility of iron in development.

Iron support spans, known for their cross section like plan, turned into a compositional sign of the nineteenth hundred years. These scaffolds included mind boggling ironwork that gave both strength and stylish allure. The Brooklyn Extension, planned by John A. Roebling and finished in 1883, embodies the capability of iron in making huge, sturdy, and outwardly striking designs.

Iron extensions have not exclusively been imperative for transportation however have likewise added to the social and engineering scenes of urban areas and districts. Their persevering through presence and useful excellence act as a demonstration of the significant effect of iron on the fabricated climate.

From Iron to Steel: A Change in Development Materials

While iron assumed a focal part in early transportation framework, the change to steel addressed a critical headway. Steel, a combination of iron, offers more prominent strength and sturdiness, settling on it an optimal decision for building extensions, high rises, and other transportation-related structures.

The late nineteenth and mid twentieth hundreds of years saw the ascent of steel spans, celebrated for their proficiency, strength, and plan adaptability. The reception of steel permitted specialists to make longer ranges, expanding the adaptability and effectiveness of transportation organizations.

The Forward Scaffold in Scotland, finished in 1890, is a famous steel span that grandstands the capability of this material. The extension's three cantilevered ranges, each with a focal range of 520 meters (1,710 feet), were built utilizing 54,000 tons of steel. This designing accomplishment featured the capacities of steel in making enormous, tough, and outwardly noteworthy designs.

The improvement of steel-supported concrete, a creative blend of steel and concrete, further extended the potential outcomes in transportation framework. This composite material offers both the strength of steel and the pliability of cement, settling on it an optimal decision for building extensions, passages, and interstates.

The Auto Upheaval: Iron and the Advanced Vehicle

Iron keeps on being a fundamental material in the auto business. The improvement of the vehicle in the late nineteenth and mid twentieth hundreds of years addressed a vital second in transportation, and iron assumed a critical part in the development of auto plan and assembling.

Early autos highlighted iron and steel parts for their body and frame. These materials gave the essential strength and inflexibility to endure the requests of street travel. The presentation of sequential construction system creation methods, strikingly spearheaded by Henry Passage, sped up the large scale manufacturing of vehicles, making them open to a more extensive populace.

As auto innovation kept on propelling, iron and steel parts became lighter and more solid, adding to the advancement of more secure and more effective vehicles. Advancements in metallurgy and assembling procedures prompted the formation of high-strength prepares that better vehicle crash execution and eco-friendliness.

Present day autos are outfitted with different security highlights, including airbags, safety belt pretensioners, and fold zones, a large number of which are produced using iron and steel. These elements have fundamentally upgraded traveler wellbeing and diminished the seriousness of auto collisions.

4.2 The Expansion of Railway Networks

The extension of railroad networks has been perhaps of the most extraordinary advancement throughout the entire existence of transportation. Throughout recent hundreds of years, railroads play had a critical impact in interfacing mainlands, forming economies, and cultivating social trades. This paper investigates the advancement of rail line frameworks, their effect on society, and their job in encouraging globalization.

The Good 'ol Days: Birth of the Railroad

The introduction of the cutting edge railroad can be followed back to the mid nineteenth hundred years, when the world saw the rise of the primary steam-controlled trains and rail line organizations. The creation of the steam motor, especially the train, made ready for another period in transportation.

The Stockton and Darlington Rail route, initiated in 1825 in Britain, is in many cases considered the world's most memorable public rail line to utilize steam trains. This railroad denoted a critical defining moment in transportation history, as it showed the capability of rail routes to convey travelers and freight productively over significant distances.

The outcome of the Stockton and Darlington Rail route prompted a quick extension of rail route networks in the Unified Realm and, in this manner, in Europe and North America. Early railroads principally centered around interfacing modern regions, mines, and ports to work with the development of natural substances and completed products.

Iron and Steam: The Structure Blocks of Development

The extension of rail route networks was made conceivable by the imaginative utilization of iron and the improvement of productive steam trains. Iron was picked for rail route tracks and scaffolds because of its solidarity and sturdiness. These early rail routes denoted the progress from wooden tracks and pony attracted vehicles to press tracks and strong steam trains.

The advancement of the steam train was driven by engineers like George Stephenson, who is frequently viewed as the "Father of Railroads." Stephenson's train, the Rocket, planned in 1829, accomplished momentous speed and effectiveness, setting new principles for rail line travel.

Iron tracks decreased contact, empowering trains to move effortlessly and productivity. Iron scaffolds, like the famous Iron Extension in Shropshire, Britain, finished in 1779, showed the plausibility of involving iron designs in railroad development. These mechanical progressions made railroads a financially savvy and proficient method of transportation.

Railroads and the Modern Upheaval

The development of rail line networks matched with the Modern Transformation, and the two peculiarities were firmly interlaced. Rail routes assumed a fundamental part in working with the transportation of unrefined components and completed merchandise, interfacing modern centers and growing business sectors.

Railroads gave makers a dependable method for moving their items to more extensive business sectors, decreasing transportation expenses, and cultivating financial development. The capacity to move merchandise quickly and proficiently by means of rail routes added to the progress of enterprises going from materials to large equipment.

Besides, the development of rail routes prompted the foundation of new businesses taking special care of the necessities of the rail line framework itself. Steel creation, for instance, saw huge extension as rail lines requested tremendous amounts of steel for tracks, scaffolds, and moving stock. This cooperative connection among rail routes and industrialization sped up financial advancement in numerous locales.

The American West: A Rail route Upheaval

In the US, the extension of rail route networks assumed a vital part in the improvement of the American West. The consummation of the Principal Crosscountry Railroad in 1869, associating the east and west drifts, was a milestone accomplishment that changed the country's scene and economy.

The Association Pacific and Focal Pacific Rail lines were answerable for building the Primary Cross-country Railroad. Their development groups, comprising of workers from assorted foundations, worked enthusiastically to lay tracks, fabricate scaffolds, and shoot burrows through the difficult territory of the western US.

The driving of the Brilliant Spike at Projection Culmination, Utah, represented the fruition of this aggressive undertaking, denoting the unification of the US and the groundbreaking force of railways. The Primary Cross-country Railroad decreased crosscountry make a trip time from months to only days, encouraging the toward the west development of pioneers and empowering the productive development of individuals and merchandise.

The extension of rail line networks in the American West prompted the development of new urban areas and ventures, including mining, agribusiness, and assembling. Rail lines opened up immense domains for settlement and doubledealing, changing the financial and social scene of the district.

Imperialism and Domain Building

The extension of rail route networks was not restricted to Europe and North America. During the nineteenth and mid twentieth hundreds of years, European powers involved rail lines for the purpose of attesting command over settlements and growing their domains.

Frontier powers saw rail routes as a device for monetary double-dealing and regulatory control. Railroads were built to work with the extraction of assets, like minerals and horticultural items, from settlements and to ship them to beach front ports for send out. The development of rail routes in pilgrim domains frequently elaborate constrained work and the uprooting of neighborhood networks.

One remarkable illustration of frontier railroad development is the English assembled rail lines in India. The tremendous Indian rail line organization, which started during the nineteenth 100 years, was instrumental in associating various locales of the Indian subcontinent, from the Himalayas toward the southern tip. The rail lines filled the double need of working with asset extraction and empowering the development of English soldiers for authoritative and vital purposes.

Likewise, in Africa, pilgrim powers constructed railroads to get to the mainland's abundance of regular assets, including minerals, elastic, and lumber. These rail routes assumed a critical part in the double-dealing of Africa's assets and the combination of pilgrim rule.

Globalization and Social Trade

The development of rail line networks significantly affected social trade and globalization. Rail routes made it workable for individuals and thoughts to travel all the more rapidly and effectively across tremendous distances, cultivating multifaceted associations and trades.

One of the most famous instances of social trade worked with by rail lines is the Orient Express. The Orient Express was an extravagance train administration that ran from Paris to Istanbul, interfacing Western Europe toward the Eastern Mediterranean and the Close to East. The train administration, which started in the late nineteenth 100 years, became inseparable from class and experience, offering explorers a novel chance to encounter various societies and scenes.

The Trans-Siberian Rail route, traversing almost 6,000 miles across Russia, is one more astounding illustration of the job of rail routes in social trade. Finished in the mid twentieth hundred years, this rail line associated Moscow to the far eastern districts of Russia, including Vladivostok. It worked with the development of individuals and merchandise across the tremendous region of Siberia, associating different ethnic gatherings and cultivating social trade.

In the US, the development of rail line networks assumed a critical part in the development of individuals and thoughts. The railroad took into consideration the scattering of writing, workmanship, and amusement, as well as the trading of social practices and customs between various locales of the country. It likewise assumed a vital part in the toward the west relocation of different gatherings, adding to the multicultural texture of the country.

Mechanical Progressions: The Period of Steam

The extension of rail route networks was firmly connected to mechanical headways in train plan and rail route foundation. The time of steam velocity saw constant enhancements in the power, effectiveness, and security of trains.

The advancement of bigger and all the more remarkable trains, frequently highlighting various driving wheels, empowered trains to pull heavier loads and navigate more extreme inclinations. The presentation of new advances, for example, the air powered brake, which considered quicker and more secure halting of trains, upgraded the wellbeing and unwavering quality of rail travel.

The utilization of iron and steel in rail line development further added to the mechanical headways of the time. Steel rails, which supplanted iron tracks, were more tough and had a more extended life expectancy, diminishing support costs and upgrading the general effectiveness of rail route frameworks.

Jolt and Fast Rail

As the twentieth century advanced, rail routes saw critical headways with the presentation of charge and the improvement of rapid rail frameworks. Jolt of railroad lines supplanted steam trains with electric foothold, offering cleaner and more effective drive.

Fast rail organizations, described by committed tracks and smoothed out trains, re-imagined rail go in the mid to late twentieth 100 years. The Shinkansen, or "projectile train," in Japan was one of the main rapid rail frameworks, setting new guidelines for speed and effectiveness. These trains, with their smooth plans and state of the art innovation, went at paces of as much as 320 kilometers each hour (200 miles each hour), changing really long travel and offering an all the more harmless to the ecosystem choice to air travel.

The progress of high velocity rail frameworks in Japan roused the advancement of comparable organizations in Europe and different regions of the planet. These frameworks were supported by progresses in designing, materials, and streamlined features, featuring the continuous development of railroads as a method of transportation.

4.3 Iron Ships and Their Impact on Global Trade

The coming of iron boats in the nineteenth century denoted an extraordinary crossroads throughout the entire existence of oceanic transportation. These vessels, built from iron and later steel, changed worldwide exchange by offering expanded productivity, dependability, and limit. This paper dives into the advancement of iron and steel sends, their effect on worldwide exchange, and their getting through heritage in the realm of transportation.

The Good 'ol Days: Wooden Boats and the Progress to Press

For quite a long time, wooden boats ruled the universe of sea transportation. Ships produced using wood were the foundation of worldwide exchange, associating mainlands and working with the trading of merchandise, culture, and thoughts. In any case, wooden vessels had intrinsic limits, including helplessness to spoil, fire, and the requirement for consistent upkeep.

The progress from wooden to press ships was driven by the requirement for more solid, sturdy, and productive vessels. Iron, as a development material, offered a few benefits. It was impervious to erosion, strong, and could be formed into complex shapes. The presentation of iron shipbuilding addressed an essential change in sea innovation.

The main iron-hulled transport, the Aaron Manby, was worked in 1821. This vessel was made of iron plates bolted together and showed the capability of iron as a shipbuilding material. Nonetheless, it was the SS Extraordinary England, planned by Isambard Realm Brunel and sent off in 1843, that undeniable a critical leap forward in the development of iron boats. This vessel was not just the principal iron-hulled, screw-impelled transport yet in addition the first with a twofold body plan.

The Incomparable England's imaginative plan added to its prosperity. The mix of an iron frame and a screw propeller, which supplanted customary paddlewheels, sped up and proficiency. The vessel's plan exhibited the capability of iron in sea transportation, making way for the eventual fate of shipbuilding.

The Change to Steel Boats

While iron boats addressed an outstanding progression in sea innovation, the shift from iron to steel denoted one more huge jump forward. Steel, a composite of iron, offered significantly more noteworthy strength, solidness, and protection from erosion. These traits settled on steel an optimal decision for transport development, and it turned into the prevailing material in the late nineteenth and mid twentieth hundreds of years.

The utilization of steel in shipbuilding considered the formation of bigger, more vigorous vessels. Steel-hulled ships were equipped for conveying more noteworthy freight loads, which was particularly significant for significant distance shipping lanes. Their sturdiness and protection from rust broadened the functional existence of vessels, lessening upkeep costs.

One of the most notorious steel-hulled ships was the RMS Titanic, which set forth in 1912. Albeit unfortunately known for its sinking, the Titanic was a demonstration of the designing and development capacities of the time. Its steel development made it an impressive vessel, and the misfortune prompted progressions in oceanic security guidelines.

Iron and Steel Boats in Worldwide Exchange

The reception of iron and steel ships significantly affected worldwide exchange. These vessels assumed a focal part in the extension and speed increase of oceanic transportation, interfacing far off ports and working with the trading of products and assets across the world.

Expanded Limit and Effectiveness: Iron and steel ships were bigger and more open than their wooden partners. This expanded freight limit and considered the transportation of a large number of merchandise, from natural substances to made items. The productivity of these vessels added to quicker conveyance times and diminished delivery costs.

Worldwide Availability: The development of iron and steel ships cultivated worldwide network by empowering the foundation of new shipping lanes. These vessels could explore longer distances and investigate areas that had recently been trying for wooden boats. The Suez Channel, which opened in 1869, gave a more straightforward course among Europe and Asia, further upgrading worldwide exchange.

Further developed Security and Unwavering quality: Iron and steel ships were more powerful and less inclined to harm than wooden boats. This expanded wellbeing adrift, lessening the dangers related with wrecks and different catastrophes. The better unwavering quality of these vessels supported trust in oceanic exchange.

Support for Industrialization: The ascent of iron and steel shipbuilding concurred with the Modern Upset. Steel creation extended to fulfill the needs of boat development, animating monetary development. The shipbuilding business itself turned into a huge wellspring of work and monetary action.

Social Trade and Movement: Iron and steel delivers likewise assumed a part in working with social trade and migration. These vessels conveyed individuals from different foundations to new terrains, adding to the multicultural texture of countries like the US, Canada, Australia, and South America. The capacity to move individuals and products proficiently and moderately changed the world's segment and monetary scenes.

The Period of Steam and Motor Fueled Boats

The reception of steam motors for transport impetus was one more crucial advancement in oceanic transportation. Steam motors gave vessels solid and effective power sources, diminishing their reliance on wind and muscle.

Steam motors fueled the original of iron and steel ships, empowering them to go against winning breezes and flows. The presentation of screw propellers, which supplanted paddlewheels, further developed drive productivity, making steamships the prevailing type of sea transportation.

One of the most renowned steamships was the RMS Sovereign Mary, which started administration during the twentieth 100 years. The Sovereign Mary, made of steel, was known for its extravagance and speed. It worked as an overseas liner and assumed a pivotal part during The Second Great War as a troopship.

The change from steam motors to gas powered motors and gas turbines further altered sea transportation. These motors were more minimal and productive, lessening the requirement for broad hardware and taking into consideration more freight space. Diesel-fueled ships, which became pervasive during the twentieth hundred years, offered significantly more prominent proficiency, dependability, and reach.

The Effect on Worldwide Shipping lanes

Iron and steel ships, fueled by steam and later by gas powered motors, reshaped worldwide shipping lanes. The capacity to travel significant distances effectively and explore through various weather patterns affected the examples of global trade and the areas of significant ports.

Transoceanic Exchange: Iron and steel ships assumed a huge part in overseas exchange, interfacing Europe and North America. These vessels gave a dependable method for shipping products and travelers across the Atlantic Sea, adding to the financial improvement of the two mainlands.

Pacific Courses: The development of exchange the Pacific Sea was one more significant result of iron and steel transport innovation. Vessels were currently equipped for heading out from North America to Asia, working with the trading of merchandise and social impacts between locales. The development of the Panama Trench in the mid twentieth century further improved this availability.

Pioneer Exchange: Iron and steel ships were instrumental in working with the development of products between pilgrim powers and their settlements. These vessels moved natural substances from settlements to royal focuses and took completed items back to pioneer markets. The expanded productivity of these boats assisted the abuse of frontier assets.

Development of Containerization: The mid-twentieth century saw the ascent of containerization, a transportation upset that further upgraded worldwide exchange. Normalized transporting holders could be handily stacked and dumped from boats, trains, and trucks. This advancement altogether diminished freight taking care of times and expenses.

The Titanic and Sea Wellbeing

The RMS Titanic, perhaps of the most well known transport ever, fills in as a sign of the significance of sea security. The sinking of the Titanic in 1912, because of a crash with a chunk of ice, brought about the deficiency of north of 1,500 lives and significantly affected the transportation business and sea guidelines.

The misfortune prompted tremendous changes in sea security practices and guidelines. Worldwide shows, like the Wellbeing of Life Adrift (SOLAS) deal, were laid out to upgrade the security of boats and travelers. These guidelines covered regions like boat development, raft necessities, and navigational methods.

The sinking of the Titanic featured the requirement for better correspondence and route frameworks, as well as the significance of rafts and life-saving gear. Wellbeing estimates executed directly following the fiasco have altogether decreased the dangers related with sea travel.

The Getting through Tradition of Iron and Steel Boats

The tradition of iron and steel ships lives on in the realm of transportation. Current freight and traveler vessels keep on being developed from steel, profiting from its solidarity, sturdiness, and protection from consumption. The utilization of cutting edge materials, including high-strength steel combinations and composites, has additionally worked on the productivity and wellbeing of boats.

Containerization, presented during the twentieth 100 years, has changed the worldwide transportation industry. Compartment ships, which can convey large number of holders, have turned into the workhorses of worldwide exchange, offering savvy and dependable vehicle of merchandise.

The change to all the more harmless to the ecosystem impetus strategies, like condensed gaseous petrol (LNG) and discharge diminishing advancements, mirrors the business' obligation to manageability. These advancements intend to diminish the ecological effect of delivery and advance cleaner sea transportation.

Chapter 5

Iron and Warfare

Iron has for some time been entwined with the historical backdrop of fighting, filling in as a basic component in the improvement of weapons, reinforcement, and devices that have molded the course of human clash. The advancement of iron's part in fighting traverses centuries, reflecting changes in innovation, methodology, and the actual idea of outfitted struggle. In this story, we will investigate the excursion of iron as a vital resource in the battlefield, following its extraordinary effect on fights and social orders from old times to the cutting edge period.

The story starts in the fogs of days of yore, where early human developments previously tackled the force of iron. Iron, which is bountiful in nature, was at first used for everyday purposes like devices and farming executes. Its solidarity and sturdiness pursued it a characteristic decision for the people who looked to develop the land and fabricate the groundworks of human progress. As social orders progressed, so did how they might interpret the likely military uses of iron.

The Iron Age, a period generally traversing from around 1200 BC to 600 BC, denoted a huge defining moment throughout the entire existence of fighting. It was during this time that iron started to supplant bronze as the essential material for making weapons and covering. The vital benefit of iron was its boundless accessibility, empowering the creation of weaponry on a lot bigger scope. This shift had sweeping ramifications for how fights were battled and the results they created.

One of the most notorious iron weapons from the old world was the iron blade. These weapons were more honed, more strong, and simpler to efficiently manufacture than their bronze ancestors. The iron blade turned into an image of force and renown for old fighters. Specifically, the Roman gladius, a short and blade that cuts both ways, would become inseparable from the discipline and expertise of the Roman armies.

Covering likewise went through huge changes in the Iron Age. Iron took into consideration the production of more strong and defensive protective layer, for example, the lorica segmentata, a sort of fragmented plate reinforcement utilized by the Roman armies. The coming of iron reinforcement denoted a basic change yet to be determined among offense and protection on the front line.

The consolidation of iron into weaponry and protective layer implied that fights could be battled on a more fabulous scale. Armed forces furnished with iron weapons and protective layer enjoyed an impressive upper hand over those actually utilizing bronze or even stone. The innovation of the time impacted the results of fights as well as the methodologies utilized by military authorities.

As iron turned out to be more pervasive, military strategies developed to take advantage of the upsides of this new material. Iron-tipped lances and bolts, for example, could puncture through foe defensive layer, while iron caps and safeguards gave prevalent security. This required a shift towards additional coordinated developments and trained battle procedures. Phalanx developments, which expected fighters to stand intently along with covering safeguards, turned into a sign of iron-age fighting.

Also, the expanded accessibility of iron prompted the rise of expert militaries. These troopers were exceptional with iron weapons and covering, frequently given by the state. The Roman armies, with their normalized gear and thorough preparation, were perfect representations of this pattern. The ascent of expert militaries denoted a critical takeoff from the dependence on resident fighters that portrayed before periods.

The effect of iron stretched out past the war zone. It assumed a significant part in the ascent and fall of domains and realms. The capacity to control iron creation and circulation turned into a wellspring of influence and abundance for some social orders. This control permitted states to keep up with standing militaries, vanquish new domains, and apply predominance over adjoining districts.

The Hittites, for instance, were early bosses of iron creation and utilized this benefit to extend their impact in the antiquated Close to East. Their iron weapons and protective layer gave them a considerable edge over their opponents. Additionally, the Assyrians, with their iron weaponry and efficient military, laid out perhaps the earliest obvious domain ever.

The meaning of iron in fighting can likewise be found in the mechanical developments it prodded. Smiths and metallurgists of the time persistently attempted to work on the quality and viability of iron weapons and defensive layer. This drive for advancement prompted the improvement of new weapon plans and manufacturing methods. Ironworkers explored different avenues regarding different amalgams and intensity therapies to make cutting edges that were sharp as well as strong.

Quite possibly of the most renowned development in ironworking was the pot steel. Created in India around the third century BC, pot steel was a top notch steel delivered by dissolving iron and different materials together. The outcome was an edge with remarkable sharpness and sturdiness, frequently alluded to as "Damascus steel." These sharp edges were profoundly pursued and became unbelievable for their cutting power.

The far reaching utilization of iron in fighting likewise had social and representative ramifications. Iron was related with strength and authority in numerous antiquated social orders. The expression "iron-fisted rule" conveys the possibility of a solid, unfaltering power. Iron was much of the time utilized in the development of fortresses and attack motors, representing the unyielding will to overcome and safeguard. Iron's emblematic weight could be felt on the combat zone as well as in the courts of lords and rulers.

The utilization of iron in fighting was not restricted to the Old World. In Mesoamerica, the Aztecs, and other pre-Columbian civic establishments, utilized ironwood clubs and obsidian sharp edges in battle. While false iron, these weapons exhibited the craving to saddle the benefits of this material, even in districts where local iron was scant. The journey for iron or iron-like materials has been a common subject throughout the entire existence of fighting.

As the hundreds of years passed, the significance of iron in fighting kept on developing. Archaic Europe saw the ascent of iron-clad knights, completely reinforced champions who ruled the combat zones of the Medieval times. These knights, furnished with swords, spears, and other iron weapons, were the encapsulation of valor and military ability. The strength and versatility of their iron protective layer were in many cases the way in to their endurance on the war zone.

The middle age time frame additionally saw huge improvements in attack fighting. Iron assumed a focal part in the development of considerable fortresses and attack motors. Palaces, with their iron-supported entryways and pinnacles, turned out to be almost invulnerable. Armed forces put vigorously in growing strong catapults and other attack motors that could break these safeguards.

The Mongol Domain, under the authority of Genghis Khan and his replacements, extended quickly across Asia and Europe, thanks to a limited extent to their gifted utilization of iron weaponry. Mongol fighters used iron-tipped bolts and blades, and their cavalry was a considerable power on the front line. Their portability and iron weapons permitted them to strike quickly and successfully against their foes.

The time of investigation in the fifteenth and sixteenth hundreds of years likewise significantly affected the historical backdrop of iron and fighting. As European powers looked for new shipping lanes and regions, they carried with them iron weapons and protection. This had wrecking ramifications for native people groups in the Americas, Africa, and different districts, who were in many cases unprepared to guard themselves against the unrivaled innovation of the European trespassers.

The presentation of guns denoted a critical change in the utilization of iron in fighting. The early guns were made of iron and utilized black powder to impel shots. The arquebus, a forerunner to the rifle, was perhaps the earliest handheld gun and assumed a urgent part in European struggle during the late middle age and early current period.

The far and wide reception of guns changed fighting. Iron cannons, which could convey destroying capability, assumed a critical part in the attack of strongholds. The improvement of flintlocks and rifles, likewise made of iron, changed the elements of infantry battle. The capability of black powder gun furnished infantry units delivered conventional reinforcement less compelling, prompting the decay of vigorously defensively covered knights.

The seventeenth century saw the rise of profoundly focused and very much equipped standing militaries in Europe. These expert warriors were furnished with iron rifles and pikes, and they framed the foundation of European military power during this period. The Ironside infantry of the English Nationwide conflict and the Swedish Caroleans of the Incomparable Northern Conflict were imposing instances of these exceptionally prepared powers.

The Napoleonic Conflicts, battled in the late eighteenth and mid nineteenth hundreds of years, saw the continuation of iron's focal job in fighting. Iron cannonballs and gunnery pieces were instrumental in the huge scope clashes of this period. Infantry actually depended on rifles and knifes, and cavalry units utilized iron sabers and spears. Napoleon's Grande Armée, furnished with the most recent iron weaponry, cut out a realm across Europe.

The American Nationwide conflict in the nineteenth century saw the following significant change in iron and fighting. The utilization of ironclad warships, for example, the USS Screen and the CSS Virginia, denoted a defining moment in maritime fighting. These iron-plated vessels were almost impenetrable to conventional wooden cannonballs, prompting the modernization of maritime armadas around the world.

Ashore, the Nationwide conflict additionally saw the broad utilization of iron as rifled flintlocks, big guns, and fortresses. The ironclad warship turned into an image of the changing idea of fighting, as innovation and development assumed an undeniably basic part in military technique.

The late nineteenth century carried further innovative progressions with the development of smokeless powder and rehashing guns. Smokeless powder worked on the proficiency and precision of guns, while rehashing rifles and machine.

5.1 Iron's Evolution in Weaponry

All through the records of mankind's set of experiences, iron has assumed a focal and groundbreaking part in the improvement of weaponry. The development of iron with regards to fighting is an account of advancement, versatility, and flexibility. From the beginning of iron utilization to its cutting edge applications in cutting edge military innovation, iron has been a main impetus behind the movement of weaponry.

The story starts in the far off past, during the Iron Age, a period when iron superseded bronze as the essential material for making weapons. Iron, plentiful in nature and simple to acquire through refining and producing, offered particular benefits over its ancestor, bronze. This change denoted a defining moment throughout the entire existence of weapons and fighting, as iron's predominant characteristics took into consideration the creation of additional viable and broadly accessible arms.

The rise of the iron sword was a pivotal turning point in the early reception of iron for military purposes. These swords were more keen, more sturdy, and more promptly producible than their bronze partners. Iron blades became images of force and notoriety, and they addressed a critical jump in weapon innovation. Prominently, the Roman gladius, a short and twofold edged iron sword, became symbolic of the Roman armies and their restrained military ability.

Defensive layer, as well, went through a huge change with the coming of iron. Iron took into consideration the making of more vigorous and defensive shield, for example, the lorica segmentata, a kind of fragmented plate covering utilized by the Roman armies. The presentation of iron reinforcement changed the harmony among offense and safeguard on the war zone, prompting shifts in military system and strategies.

The ascent of iron weaponry prompted more coordinated and restrained battle strategies. Iron-tipped lances and bolts, specifically, could puncture foe reinforcement, requiring the utilization of closely knit arrangements like the phalanx. This undeniable a takeoff from the more individualistic battle styles of prior times and laid the foundation for the improvement of expert militaries.

One of the main results of iron's far and wide reception in weaponry was the development of expert militaries. These standing armed forces, outfitted with normalized iron weapons and protection, were a noticeable takeoff from the dependence on resident fighters that portrayed before periods. The Roman armies, known for their severe discipline and exceptional fighters, exemplified the benefits of expert powers.

The utilization of iron in weaponry likewise had significant ramifications for the power and impact of social orders. The capacity to control iron creation and circulation turned into a wellspring of solidarity and abundance for some states. The Hittites, who were early bosses of iron creation, utilized their iron weapons to grow their impact in the antiquated Close to East. Essentially, the Assyrians, with their iron weaponry and efficient military, laid out quite possibly the earliest evident domain ever.

Advancement in ironworking was one more sign of this period. Smiths and metallurgists consistently tried to work on the quality and viability of iron weapons and protective layer. These craftsmans tried different things with various compounds and intensity therapies to make cutting edges that were sharp as well as versatile. Damascus steel, well known for its unmistakable wavy examples and uncommon sharpness, was a result of such trial and error and craftsmanship.

The meaning of iron in fighting reached out to its social and emblematic ramifications. Iron became related with strength and authority in numerous old social orders. Phrases like "iron-fisted rule" convey the possibility of a solid, unwavering power. Iron was in many cases utilized in the development of fortresses and attack motors, representing the unflinching will to vanquish and safeguard. Iron's emblematic weight was felt on the combat zone as well as in the courts of lords and rulers.

Iron's job in fighting was not bound to the Old World. In Mesoamerica, the Aztecs and other pre-Columbian human advancements utilized ironwood clubs and obsidian edges in battle. While these weapons didn't consolidate genuine iron, they exhibited the longing to saddle the benefits of this material, even in locales where local iron was scant. The quest for iron or iron-like materials turned into a repetitive topic throughout the entire existence of weaponry.

As the hundreds of years passed, the significance of iron in fighting kept on developing. Archaic Europe saw the development of iron-clad knights, completely shielded champions who overwhelmed the war zones of the Medieval times. These knights, furnished with blades, spears, and other iron weapons, exemplified valor and military ability. The strength and versatility of their iron protective layer frequently resolved their endurance in fight.

Middle age fighting additionally saw huge progressions in attack innovation. Iron assumed a focal part in the development of imposing strongholds and attack motors. Palaces, sustained with iron-built up doors and pinnacles, turned out to be almost secure. Armed forces put vigorously in growing strong catapults and other attack motors to penetrate these protections.

The Mongol Realm, under the authority of Genghis Khan and his replacements, extended quickly across Asia and Europe, thanks to a limited extent to their talented utilization of iron weaponry. Mongol champions used iron-tipped bolts and blades, and their rangers, furnished with iron sabers and spears, was a considerable power on the front line. Their versatility and iron weapons permitted them to strike quickly and actually against their foes.

The time of investigation in the fifteenth and sixteenth hundreds of years likewise significantly affected the historical backdrop of iron and weaponry. As European powers looked for new shipping lanes and domains, they acquainted iron weapons and reinforcement with native people groups in the Americas, Africa, and different areas. These native populaces were in many cases unfit to safeguard themselves against the prevalent innovation of the European trespassers, which included iron-based weaponry.

The presentation of guns denoted a critical change in the utilization of iron in weaponry. Early guns were made of iron and utilized black powder to move shots. The arquebus, a forerunner to the black powder rifle, was perhaps the earliest handheld gun and assumed a significant part in European clash during the late middle age and early current period.

The far and wide reception of guns upset fighting. Iron cannons, which could convey destroying capability, assumed a significant part in the attack of strongholds.

The advancement of guns and rifles, additionally made of iron, changed the elements of infantry battle. The capability of gun outfitted infantry units delivered conventional reinforcement less viable, prompting the downfall of vigorously defensively covered knights.

The seventeenth century saw the development of profoundly focused and very much outfitted standing armed forces in Europe. These expert warriors were furnished with iron black powder guns and pikes, shaping the foundation of European military power during this period. The Ironside infantry of the English Nationwide conflict and the Swedish Caroleans of the Incomparable Northern Conflict were considerable instances of these exceptionally prepared powers.

The Napoleonic Conflicts in the late eighteenth and mid nineteenth hundreds of years kept on highlighting the focal job of iron in weaponry. Iron cannonballs and big guns pieces were instrumental in the huge scope skirmishes of this time. Infantry actually depended on rifles and knifes, while rangers units utilized iron sabers and spears. Napoleon's Grande Armée, furnished with the most recent iron weaponry, cut out a realm across Europe.

The American Nationwide conflict in the nineteenth century denoted one more huge achievement in the advancement of iron in weaponry. The utilization of ironclad warships, for example, the USS Screen and the CSS Virginia, addressed a defining moment in maritime fighting. These iron-plated vessels were almost impenetrable to customary wooden cannonballs, prompting the modernization of maritime armadas around the world.

Ashore, the Nationwide conflict likewise saw the broad utilization of iron as rifled flintlocks, big guns, and strongholds. The ironclad warship turned into an image of the changing idea of fighting, as innovation and development assumed an undeniably basic part in military system.

The late nineteenth century carried further innovative headways with the development of smokeless powder and rehashing guns. Smokeless powder worked on the effectiveness and precision of guns, while rehashing rifles and automatic weapons emphatically expanded the pace of shoot. These developments changed infantry strategies and made guard really testing.

The Second Great War, known as the Incomparable Conflict, exhibited the staggering force of iron and steel in weaponry. Iron and steel were utilized in many weapons, from ordnance and assault rifles to tanks and spiked metal. The merciless close quarters conflict of The Second Great War, where troopers confronted steady dangers from iron and steel, featured the terrible truth of industrialized fighting.

The presentation of tanks in The Second Great War addressed another critical achievement. These defensively covered vehicles, outfitted with iron plating and weighty weapons, gave a better approach to get through foe lines. The tank, at first an English development, turned into an essential part of current defensively covered fighting.

The Second Great War saw a further development of iron in weaponry. Tanks turned out to be more impressive with thicker covering and all the more remarkable firearms. Airplane, frequently developed with aluminum and different metals, assumed a definitive part in the contention. Iron stayed urgent in the creation of guns, gunnery, and the foundation of war, including extensions and fortresses.

The improvement of the nuclear bomb, which tackled the force of atomic responses, addressed another degree of damaging power during the twentieth 100 years. While the actual bomb wasn't made of iron, the designs and instruments that conveyed it frequently were. The bombings of Hiroshima and Nagasaki in 1945 achieved the finish of The Second Great War and another period .

5.2 The Iron Age and Its Military Implications

The Iron Age, a urgent period in mankind's set of experiences, was set apart by the far reaching reception of iron as the essential material for making devices, weapons, and reinforcement. This age, which followed the Bronze Age, included a time of significant mechanical and cultural change. The appearance of iron had expansive ramifications for military undertakings, rethinking the idea of fighting, the systems utilized, and the power elements among social orders.

Iron's excursion as an imperative asset started in the old world, and its utilization in weaponry addresses a huge change throughout the entire existence of contention. The Iron Age, comprehensively dated from around 1200 BC to 600 BC, started a significant change in how wars were battled. The essential benefit of iron over bronze was its overflow in nature, making it all the more promptly accessible and reasonable for the large scale manufacturing of weapons, defensive layer, and apparatuses.

One of the most notorious utilizations of iron in the old military setting was the iron blade. These blades, contrasted with their bronze ancestors, offered unrivaled sharpness, solidness, and simplicity of creation. Iron blades turned into an image of force and a proportion of glory among old fighters. Strikingly, the Roman gladius, a twofold edged and short iron blade, became symbolic of the discipline and military expertise of the Roman armies.

Covering, as well, went through a huge development during the Iron Age. The presentation of iron took into consideration the formation of more hearty and defensive protective layer, for example, the lorica segmentata, a sectioned plate reinforcement generally utilized by the Roman armies. The execution of iron reinforcement in a general sense moved the harmony among offense and safeguard on the war zone, catalyzing changes in military systems.

The ascent of iron weapons required the improvement of more coordinated and trained battle methods. Iron-tipped lances and bolts, with their ability to infiltrate foe reinforcement, required the utilization of closely knit arrangements like the phalanx. These developments, where fighters stood intently along with covering safeguards, denoted a takeoff from the more individualistic and disordered battle styles of before times.

One of the most significant ramifications of iron's inescapable reception was the ascent of expert militaries. These militaries, outfitted with normalized iron weapons and protective layer, connoted a huge shift from the dependence on resident officers that described before periods. The Roman armies, famous for their severe discipline, normalized gear, and thorough preparation, filled in as a model of the benefits of expert military powers.

The utilization of iron in weaponry likewise had significant repercussions for the power elements among social orders. The control of iron creation and conveyance turned into a wellspring of riches and impact for some states. The Hittites, for instance, were early bosses of iron creation, and they used their iron weaponry to grow their impact in the old Close to East. Also, the Assyrians, with their iron weapons and efficient military, laid out perhaps of the earliest known domain ever.

Advancements in ironworking were one more sign of the Iron Age. Smiths and metallurgists persistently attempted to upgrade the quality and adequacy of iron weapons and protective layer. These skilled workers tried different things with various combinations and intensity therapies to make cutting edges that were sharp as well as versatile. Damascus steel, commended for its unmistakable wavy examples and excellent sharpness, arose as a result of this trial and error and craftsmanship.

Iron's part in fighting stretched out past the front line; it had social and representative importance in numerous antiquated social orders. Iron became inseparable from strength and authority. Phrases like "iron-fisted rule" conveyed the possibility of solid, unfaltering power. Iron was habitually utilized in the development of fortresses and attack motors, representing a resolute assurance to overcome and safeguard. Iron's emblematic weight was felt on the front line as well as in the courts of rulers and heads.

Iron's impact in fighting was not confined to the Old World; it reached out to different locales also. In Mesoamerica, the Aztecs and other pre-Columbian developments utilized ironwood clubs and obsidian edges in battle. While these weapons didn't consolidate genuine iron, they uncovered a longing to outfit the upsides of this material, even in regions where local iron was scant. The quest for iron or iron-like materials turned into a common subject throughout the entire existence of fighting.

As time walked on, the meaning of iron in fighting kept on developing. Archaic Europe saw the rise of iron-clad knights, completely defensively covered heroes who ruled the front lines of the Medieval times. These knights, furnished with iron blades, spears, and different weapons, addressed the encapsulation of gallantry and military ability. The flexibility and strength of their iron defensive layer were many times the way in to their endurance on the combat zone.

Middle age fighting likewise seen critical advances in attack innovation. Iron assumed a focal part in the development of considerable strongholds and attack motors. Palaces, sustained with iron-built up entryways and pinnacles, turned out to be almost invulnerable. Armed forces put vigorously in growing strong catapults and other attack motors to break these guards.

The Mongol Realm, under the authority of Genghis Khan and his replacements, extended quickly across Asia and Europe, thanks to some extent to their gifted utilization of iron weaponry. Mongol fighters used iron-tipped bolts and blades, and their rangers, furnished with iron sabers and spears, was an impressive power on the war zone. Their versatility and iron weapons permitted them to strike quickly and successfully against their foes.

The period of investigation in the fifteenth and sixteenth hundreds of years likewise left a critical engraving on the historical backdrop of iron and fighting. As European powers looked for new shipping lanes and domains, they acquainted iron weapons and shield with native people groups in the Americas, Africa, and different locales. These native populaces were in many cases unprepared to shield themselves against the prevalent innovation of the European trespassers, which included iron-based weaponry.

The presentation of guns addressed a stupendous change in the utilization of iron in fighting. Early guns were made of iron and used black powder to push shots. The arquebus, a forerunner to the black powder gun, was quite possibly the earliest handheld gun and assumed a urgent part in European clash during the late middle age and early current period.

The far reaching reception of guns changed fighting. Iron guns, which could convey destroying capability, assumed a significant part in the attack of strongholds. The improvement of guns and rifles, likewise made of iron, changed the elements of infantry battle. The capability of rifle outfitted infantry units delivered customary defensive layer less viable, adding to the downfall of vigorously shielded knights.

The seventeenth century saw the rise of exceptionally focused and very much outfitted standing militaries in Europe. These expert warriors were outfitted with iron guns and knifes, framing the foundation of European military power during this period. The Ironside infantry of the English Nationwide conflict and the Swedish Caroleans of the Incomparable Northern Conflict filled in as considerable instances of these profoundly prepared powers.

The Napoleonic Conflicts in the late eighteenth and mid nineteenth hundreds of years kept on highlighting the focal job of iron in weaponry. Iron cannonballs and big guns pieces were instrumental in the enormous scope clashes of this time. Infantry actually depended on flintlocks and knifes, while mounted force units utilized iron sabers and spears. Napoleon's Grande Armée, furnished with the most recent iron weaponry, cut out a realm across Europe.

The American Nationwide conflict in the nineteenth century denoted one more critical achievement in the advancement of iron in weaponry. The utilization of ironclad warships, for example, the USS Screen and the CSS Virginia, addressed a defining moment in maritime fighting. These iron-plated vessels were almost impenetrable to customary wooden cannonballs, prompting the modernization of maritime armadas around the world.

Ashore, the Nationwide conflict likewise saw the broad utilization of iron as rifled guns, ordnance, and strongholds. The ironclad warship turned into an image of the changing idea of fighting, as innovation and development assumed an undeniably basic part in military methodology.

The late nineteenth century carried further mechanical progressions with the innovation of smokeless powder and rehashing guns. Smokeless powder worked on the effectiveness and precision of guns, while rehashing rifles and automatic weapons emphatically expanded the pace of shoot. These developments changed infantry strategies and made safeguard seriously testing.

The Second Great War, known as the Incomparable Conflict, displayed the staggering force of iron and steel in weaponry. Iron and steel were utilized in a large number of weapons, from cannons and automatic rifles to tanks and spiked metal. The severe close quarters conflict of The Second Great War, where fighters confronted steady dangers from iron and steel, featured the troubling truth of industrialized fighting.

The presentation of tanks in The Second Great War addressed another huge achievement. These defensively covered vehicles, furnished with iron plating and weighty weapons, gave a better approach to get through foe lines. The tank, at first an English creation, turned into an essential part of present day reinforced fighting.

The Second Great War saw a further development of iron in weaponry. Tanks turned out to be more imposing with thicker shield and all the more impressive weapons. Airplane, frequently built with aluminum and different metals, assumed an unequivocal part in the contention. Iron stayed significant in the creation of guns, ordnance, and the foundation of war, including scaffolds and fortresses.

5.3 Modern Military Technology and Iron's Ongoing Role

In the steadily developing scene of fighting, present day military innovation remains as a demonstration of human resourcefulness, development, and the persevering through job of materials like iron. Throughout the long term, iron has stayed a focal component in the improvement of military equipment, assuming a basic part in the production of weapons, vehicles, and gear that have reshaped the idea of contention. From the modern unrest to the computerized age, the account of iron's continuous job in current military innovation is one of consistent variation, change, and progression.

The Modern Transformation, which unfurled in the late eighteenth and nineteenth hundreds of years, denoted a critical crossroads throughout the entire existence of military innovation and the boundless utilization of iron. This period achieved an extreme change in assembling and creation processes, with the large scale manufacturing of iron weapons, hardware, and apparatus turning into the new standard. The accessibility of iron for a huge scope changed the capacities of present day militaries and extended their munititions stockpile.

Quite possibly of the main advancement in military innovation during this period was the improvement of rifled black powder guns. These guns included winding notches inside the barrel that granted a twist to the shot, significantly further developing precision and reach. The combination of iron in the assembling of these guns made them more solid and dependable. Rifled flintlocks, for example, the Springfield Model 1861, assumed a pivotal part in the American Nationwide conflict, where they showed the viability of iron and the advances in guns innovation.

Iron likewise tracked down broad use in cannons during the Modern Upset. Iron cannonballs, ordnance pieces, and weapon carriages became fundamental parts of military missions. The change from bronze to press in mounted guns development addressed a jump in capability and strength. Gunnery assumed a critical part in numerous nineteenth century clashes, including the Crimean War and the American Nationwide conflict, where iron-based weapons and hardware characterized the war zones.

Reinforced fighting started to come to fruition during the late nineteenth hundred years and kept on developing in the twentieth hundred years, with iron assuming a focal part. The presentation of tanks altered the combat zone. These reinforced vehicles, commonly made of iron and later steel, consolidated portability and capability to get through adversary lines. The English Imprint I tank, utilized in The Second Great War, was a fundamental illustration of early tank advancement. The sheer presence of iron protection on the combat zone changed the elements of fighting, provoking advancements in enemy of tank weaponry and strategies.

The ascent of the airplane as a tactical resource further stressed the significance of iron in present day military innovation. While the airplane's design frequently included materials like aluminum, its powerplants weaponry actually depended on iron and steel parts. Iron-based motors and assault rifles were basic components of early airplane. During The Second Great War, ironclad airplane like the Sopwith Camel and the Fokker Dr.I assumed vital parts in surveillance and battle. The utilization of iron and steel in avionics innovation extended quickly, prompting the improvement of bigger, all the more impressive airplane in the following many years.

The period between The Second Great War and The Second Great War saw proceeded with headways in military innovation, with iron as an imperative part. Tanks expanded, all the more intensely protected, and all the more remarkable, with thicker steel plating supplanting the iron defensive layer of prior ages. The period of "lightning war," described by quick heavily clad arrangements, depended on the iron and steel of tanks like the German Panzer III and the Soviet T-34. Iron-based framework, like extensions and strongholds, assumed a huge part in military coordinated operations and system during this period.

The staggering force of iron and steel in fighting was additionally highlighted by the improvement of the nuclear bomb during The Second Great War. While the actual bomb was not built of iron, the designs and components that conveyed it were frequently made out of iron and steel.

The bombings of Hiroshima and Nagasaki in 1945 achieved the finish of The Second Great War and started another period throughout the entire existence of present day military innovation.

The post-The Second Great War time frame saw a proceeding with dependence on iron and steel as the foundation of military innovation. The Virus War competition between the US and the Soviet Association prodded the advancement of ironclad conflict machines, including tanks, shielded work force transporters, and rocket frameworks. Steel plating gave assurance against customary and atomic dangers, while iron-based framework, similar to fortifications and rocket storehouses, became key parts of military system.

The late twentieth century achieved headways in materials science and innovation that extended the potential outcomes of current military innovation. Composite materials, pottery, and combinations offered options in contrast to conventional iron and steel, giving lighter yet more sturdy protective layer and weaponry. These materials added to the improvement of additional flexible and deft military stages, for example, the M1 Abrams tank, which integrated progressed composite protection.

Covertness innovation, a basic development in present day military innovation, has likewise assumed a critical part in the improvement of airplane and other military equipment. Secrecy innovation plans to decrease radar location, making military resources less helpless against foe observation and assaults. While not straightforwardly connected with iron, the reconciliation of secrecy highlights underlines the continuous significance of materials and innovation in current fighting.

Automated flying vehicles (UAVs) and drones have become unmistakable highlights of present day military innovation. These vehicles, frequently developed with lightweight materials, serve different jobs, from observation and reconnaissance to accuracy strikes. Drones have reformed military tasks, giving continuous knowledge and the ability to draw in focuses without endangering human lives. Their prosperity highlights the proceeded with meaning of materials and innovation in reshaping military systems and strategies.

The 21st century has likewise seen an expanded spotlight on digital fighting, where the fights are battled in the computerized domain. While the space of digital fighting doesn't depend on conventional iron-based materials, it is personally associated with the proceeded with reliance on iron-based PC frameworks and organizations in military activities. The security and versatility of these frameworks have become fundamental in present day military innovation.

Planning ahead, the job of iron in current military innovation is probably going to develop. Propels in materials science might prompt much stronger and lightweight covering and weaponry. The improvement of nanomaterials and progressed compounds could additionally upgrade the exhibition of military hardware.

Man-made brainpower and robotization might change the manners by which fights are battled, with automated frameworks taking on additional jobs on the combat zone. The continuous advancement of room innovation could broaden the battlefield past Earth's climate, requiring iron and different materials to endure the afflictions of room.

Chapter 6

Iron in Everyday Life

Iron, perhaps of the most bountiful component on The planet and a principal building block of current culture, assumes an unavoidable and fundamental part in our day to day existences. From the steel structures that help our structures and extensions to the iron-rich food sources that support our bodies, this flexible component is woven into the texture of our reality. This article investigates the bunch manners by which iron effects our day to day schedules, from its presence in framework and transportation to its job in nourishment and wellbeing, and dives into the authentic, modern, and mechanical parts of iron in day to day existence.

Iron is ubiquitous in the foundation that encompasses us, forming our advanced urban areas and supporting our day to day exercises. One of the most noticeable utilizations of iron is in development, where it is an essential part of steel, a flexible and strong material utilized in structures, spans, and different designs. High rises like the Domain State Working in New York City, with its steel outline, stand as famous instances of the job of iron in forming metropolitan scenes. The Brilliant Entryway Scaffold in San Francisco, an image of designing and building ability, is one more declaration to the strength and versatility of iron-based materials. Steel-supported concrete, utilized in most current structures, is made by implanting steel bars, giving the strength expected to endure both compressive and ductile powers.

The transportation area depends intensely on iron for the development of vehicles and framework. Autos, trains, boats, and planes generally integrate iron-based materials to differing degrees. In the car business, the utilization of iron and steel gives the fundamental strength and wellbeing highlights as frame, body boards, and security confines. In rail transportation, the iron rails themselves, alongside haggles parts, are fundamental for the activity of trains.

Iron assumes an essential part in the transportation business, from the development of the actual boats to the holders utilized for moving merchandise across the world's seas. Indeed, even airplane, in spite of their lightweight development utilizing materials like aluminum and composite materials, depend on iron as motors, landing gear, and other primary parts.

The development and support of streets, one more basic part of transportation framework, include the broad utilization of iron-based materials. Black-top, the essential material for street surfaces, is blended in with totals that frequently contain iron mineral. Iron and steel extensions and bridges are fundamental for exploring testing territory and associating networks. Guardrails, signs, and streetlamps, all produced using iron and steel, add to street wellbeing and proficient traffic the executives.

Iron's job in transportation stretches out to the improvement of rail lines, which play had a vital impact in the development and network of countries. Rail routes, known as the "iron pony" in the nineteenth 100 years, altered travel, exchange, and the development of products. Iron rails, built up by steel in current times, give the establishment to rail transport. Trains, carts, and other moving stock are prevalently made of iron and steel parts. The development of fast trains, which integrate progressed materials, actually depends on iron-rich tracks for strength and security.

In the sea business, iron and steel are the essential materials for transport development. The toughness and erosion obstruction of iron and steel pursue them ideal decisions for vessels that should endure the cruel states of the untamed ocean. Freight ships, luxury ships, and military vessels are totally built with iron-based materials. The utilization of iron mineral, purified into steel, additionally stretches out to the production of steel trailers, which are fundamental to the worldwide inventory network, empowering the proficient vehicle of merchandise via ocean.

Indeed, even the airplane business, known for its dependence on lightweight materials, consolidates iron in fundamental parts. Stream motors, basic to the impetus of present day airplane, incorporate iron-based compounds that can endure high temperatures and tensions. Airplane landing gear, which should bear the weight and effect of departures and arrivals, is principally made of iron and steel. The sturdiness and strength of these materials guarantee the security of air travel.

Iron likewise plays a primary job in the field of energy creation, from the development of force plants to the age of power. Iron and steel are vital in the development of force plant framework, like boilers, turbines, and generators. These materials can endure the outrageous circumstances and tensions related with the transformation of intensity into power. Electrical transformers, used to change voltage levels in the power lattice, contain iron centers that empower proficient energy transmission.

The age of power from different sources depends on iron-based advances. Coal-terminated power plants utilize iron boilers and turbines to change over the energy of consuming coal into power. Thermal energy stations additionally utilize iron and steel in their development, with the additional advantage of radiation protecting. In sustainable power areas, iron and steel are necessary to wind turbines, sunlight based charger outlines, and hydroelectric generators. These innovations saddle nature's assets to deliver clean energy while profiting from the solidness and erosion opposition of iron-based materials.

Iron's effect on daily existence reaches out to the domain of media communications and data innovation. The development of electronic gadgets, including cell phones, PCs, and servers, includes the utilization of iron and steel in different parts. The housings of numerous electronic gadgets are built from iron or steel for toughness and electromagnetic protecting. Iron-based materials are additionally utilized in the assembling of semiconductors, which are the structure blocks of current gadgets.

The worldwide broadcast communications organization, which works with the trading of data across immense distances, depends on iron as fiber optic links. These links, encased in iron or steel sheaths, safeguard the fragile glass strands that communicate information as beats of light. Iron's job in the media communications industry guarantees the fast transmission of data, associating people and organizations all over the planet.

Iron and steel keep on supporting basic framework, including water supply and disinfection frameworks. Water pipelines, answerable for the circulation of clean water to homes and organizations, are prevalently developed utilizing iron and steel materials. These materials offer the vital strength and consumption protection from endure the afflictions of shipping water. Also, sewage and wastewater treatment offices consolidate iron and steel parts to deal with the protected removal of waste and safeguard general wellbeing.

In the field of medical services, iron assumes a fundamental part in the working of the human body. Iron is a fundamental dietary mineral that the body needs for various physiological cycles, remembering the vehicle of oxygen by hemoglobin for red platelets. A lack in dietary iron can prompt pallor, a condition described by weariness, shortcoming, and diminished mental capability. Iron enhancements and iron-rich food sources, like red meat, spinach, and sustained cereals, are significant for keeping up with legitimate iron levels in the body.

Iron likewise assumes an essential part in the drug business, where it is utilized in the production of iron-based prescriptions and enhancements. Iron enhancements are recommended to people with iron-inadequacy pallor or other ailments that require iron supplementation. Iron-based prescriptions are likewise utilized in the therapy of iron over-burden problems and certain constant illnesses.

The job of iron stretches out to the universe of design, where it is utilized in the coloring and shading of materials. Iron oxide shades, frequently got from iron metal, are utilized to make a large number of varieties in dress and texture coloring. These shades are utilized in both normal and engineered colors, giving a range of tints in the style business.

The creation of family merchandise and machines vigorously depends on iron and steel. Kitchenware, including pots, dish, and utensils, is frequently built from iron or steel for its toughness and intensity leading properties. Pressing sheets and irons, utilized for article of clothing care, are fundamental things in numerous families. Pressing sheets, with iron racks made of iron or steel, give a steady surface to dress upkeep.

Security and wellbeing are improved by the utilization of iron-based materials in lock and key frameworks. Iron keys and locks have a long history of giving security to homes, organizations, and significant belongings. The utilization of iron in these frameworks is a demonstration of the solidness and dependability of this fundamental metal.

The cutting edge universe of sports and diversion likewise includes the impact of iron. Sporting gear, for example, golf clubs, homerun sticks, and bikes, frequently consolidates iron-based materials to give strength, sturdiness, and execution. Golf clubs are weighted with iron for equilibrium and power, while polished ash might incorporate iron centers to upgrade their hitting limit. In the domain of trans-portation, bikes and cruisers have iron or steel approaches that give underlying uprightness and heartiness.

Media outlets, including music and film, features iron's adaptability in different ways. Instruments like pianos and steel drums utilize iron strings or parts to deliver sound. The reverberation and strength of iron are pivotal for the nature of sound and the life expectancy of these instruments. In filmmaking, iron-based props and gear add to the formation of sensible embellishments, from blasts to on-screen fights.

6.1 Iron in Tools and Appliances

The utilization of iron in apparatuses and machines is profoundly imbued in mankind's set of experiences and current life. From the most essential hand devices to the high level hardware that controls our businesses, iron assumes a crucial part in forming our day to day schedules and mechanical advancement. This paper investigates the broad impact of iron in the improvement of devices and machines, diving into the verifiable, modern, and mechanical viewpoints that make iron a universal component in our daily existences.

Devices have been a necessary piece of human development for centuries, and iron plays had a significant impact in their advancement. From the earliest long stretches of blacksmithing to the high level machining cycles of the present, iron remaining parts a key material in device development.

The change from stone devices to press instruments denoted a huge jump in human mechanical progression. Iron devices were more keen, harder, and more tough than their stone partners. Iron's capacity to keep a sharp edge made it crucial for cutting, molding, and framing different materials. The broad utilization of iron apparatuses altered horticulture, development, and craftsmanship, taking into consideration the formation of additional complicated and productive designs and antiquities.

Blacksmithing, the specialty of manufacturing iron and steel, turned into a foundation of hardware and weapon creation. Metal forgers were fundamental individuals from networks, creating a wide cluster of devices, from sledges and etches to tomahawks and saws. These apparatuses, intended for explicit assignments, further developed efficiency and effectiveness in different exchanges and enterprises.

The development and support of framework, an essential part of human progress, depend vigorously on iron-based instruments. Development laborers use instruments like sledges, wrenches, and bores, which consolidate iron parts for toughness and strength. Iron saws and cutting instruments are fundamental for molding building materials, like wood and metal. The capacity to cut and shape materials definitively is essential in making protected and stable designs.

In the auto business, iron devices keep on being imperative for the development and support of vehicles. Mechanics depend on iron wrenches, attachment sets, and other hand devices to support vehicles and trucks. The strength and sturdiness of iron devices are fundamental for slackening screws, nuts, and different clasp that endure the afflictions of motor activity.

Iron instruments are additionally fundamental for different modern and assembling processes. Machine shops and industrial facilities utilize iron-based apparatuses, for example, machine instruments and processing cutters, to shape and machine metal parts with accuracy. Iron's hardness and protection from wear are basic credits in guaranteeing that these apparatuses can endure high velocity machining processes and keep up with their forefronts.

The improvement of iron instruments is intently attached to the headway of designing and innovation. Machine instruments, fueled by steam motors and, later, power, upset assembling processes. Machines, processing machines, and planers, which consolidate iron parts, considered the large scale manufacturing of normalized parts and items. These apparatuses were instrumental in the development of the modern area and the modernization of economies.

Iron-based apparatuses likewise track down applications in the field of agribusiness. Cultivating hardware, from furrows and harrows to cultivators and seeders, depends on iron parts for their development. These instruments are fundamental for planning soil, establishing yields, and gathering farming produce. Iron's solidarity and strength guarantee that these devices can persevere through the difficult states of the horticultural climate.

The job of iron in the development and activity of transportation vehicles stretches out to apparatuses utilized for vehicle upkeep and fix. Iron devices, like jacks, wrenches, and tire irons, are fundamental for adjusting vehicles, trucks, bikes, and bikes. These apparatuses are intended to endure the mechanical burdens and powers engaged with vehicle fix.

The development and support of rail routes, a basic piece of worldwide transportation organizations, likewise include the utilization of iron instruments. Railroad tracks, with their iron and steel parts, require particular devices for establishment and support. Iron wrenches, spike batters, and track checks are among the instruments used to guarantee the wellbeing and productivity of rail line frameworks.

The universe of flight and aviation depends on iron-based apparatuses for the gathering, upkeep, and fix of airplane and space apparatus. Mechanics and professionals utilize iron wrenches, screwdrivers, and other hand instruments to support airplane motors, landing stuff, and aeronautics frameworks. Iron's solidarity and strength are basic credits for devices utilized in the airplane business, where wellbeing and accuracy are central.

The assembling area draws upon iron apparatuses in different ways, from the machining of metal parts to the gathering of items. Machine devices, including machines, processing machines, and drill presses, are fundamental for the production of normalized parts utilized in enterprises like auto assembling, aviation, and gadgets. Iron devices work with the creation of many-sided and complex parts with accuracy.

The field of development, which includes the structure of designs and framework, depends vigorously on iron apparatuses. Iron mallets, saws, and estimating apparatuses are fundamental for errands going from outlining and material to substantial work and wrapping up. The strength and toughness of iron apparatuses are basic for the culmination of development projects that require accuracy and security.

In the domain of carpentry and carpentry, iron apparatuses are vital for molding, cutting, and collecting wood. Hand saws, planes, etches, and cinches frequently consolidate iron parts for strength and accuracy. Iron's capacity to keep a sharp edge is fundamental for making smooth and exact cuts in wood, fundamental for furniture making and cabinetry.

Iron's job in the development of transportation foundation, including streets and scaffolds, reaches out to the utilization of iron apparatuses in these tasks. Iron wrenches, bolt cutters, and other hand instruments are fundamental for the gathering and support of extensions, bridges, and streets. The capacity to fix bolts, cut materials, and make changes guarantees the wellbeing and trustworthiness of transportation organizations.

The utilization of iron apparatuses in the field of digging is basic for removing minerals and assets from the Earth. Mining activities depend on instruments like picks, digging tools, and penetrates, frequently made with iron parts, to get through rock and soil. Iron's sturdiness is vital for enduring the rough and testing conditions experienced in mining.

The rural area, fundamental for food creation and food, benefits from the utilization of iron devices in cultivating and development. Cultivating executes like furrows, harrows, and cultivators are frequently built with iron parts for their sturdiness and strength. These instruments are utilized to get ready soil, plant crops, and oversee rural land, guaranteeing effective food creation.

The development and support of rail lines, a basic piece of worldwide transportation organizations, likewise include the utilization of iron instruments. Railroad tracks, with their iron and steel parts, require specific devices for establishment and support. Iron wrenches, spike destroys, and track checks are among the instruments used to guarantee the wellbeing and effectiveness of rail line frameworks.

The avionics and aviation businesses, known for their dependence on accuracy and wellbeing, use iron instruments for airplane get together, upkeep, and fix. Mechanics and experts utilize iron wrenches, screwdrivers, and other hand devices to support airplane motors, landing stuff, and flying frameworks. Iron's solidarity and toughness are basic ascribes for apparatuses utilized in the aeronautic trade, where wellbeing and accuracy are principal.

The domain of carpentry and carpentry, vital to the development of designs and furniture, depends on iron apparatuses for molding and gathering. Hand saws, planes, etches, and clips frequently integrate iron parts for their solidness and accuracy. Iron's capacity to keep a sharp edge is imperative for making smooth and exact cuts in wood, fundamental for furniture making and cabinetry.

The development of transportation framework, including streets and extensions, requires the utilization of iron apparatuses. Iron wrenches, bolt cutters, and other hand instruments are fundamental for the get together and support of extensions, bridges, and streets. The capacity to fix bolts, cut materials, and make changes guarantees the wellbeing and uprightness of transportation organizations.

Mining, an essential industry for the extraction of minerals and assets, draws upon the strength and solidness of iron devices. Picks, digging tools, drills, and other mining devices frequently integrate iron parts for their sturdiness and flexibility. These apparatuses are fundamental for getting through rock and soil to get to important minerals and materials.

The field of horticulture, liable for food creation and food, depends vigorously on iron apparatuses in cultivating and development. Cultivating executes like furrows, harrows, and cultivators are frequently built with iron parts for their solidness and strength. These apparatuses are utilized to get ready soil, plant crops, and oversee agrarian land, guaranteeing effective food creation.

As innovation progressed, so did the intricacy and accuracy of devices. The improvement of machine devices in the eighteenth and nineteenth hundreds of years achieved an upheaval in assembling. Machine apparatuses, like machines and processing machines, used iron parts for their development, and these devices became instrumental in the creation of normalized and tradable parts.

6.2 Iron's Contributions to Modern Living

Iron, a bountiful and adaptable component, has made a permanent imprint on the advanced world, enhancing our day to day routines in various ways. From its part in framework and innovation to its impact on medical care and the climate, iron's commitments to present day living are unavoidable and significant. This article investigates the complex effect of iron in contemporary society, enveloping verifiable, modern, and mechanical angles that support its importance.

Iron's importance is conspicuously shown in the foundation that encompasses us, forming present day urban areas and supporting everyday exercises. One of the most apparent utilizations of iron is in development, where it shapes the foundation of the fabricated climate. Steel, a flexible and hearty compound of iron, is a key material utilized in the development of high rises, spans, and other fundamental designs. Notable milestones like the Domain State Working in New York City, with its steel outline, stand as demonstration of the job of iron in forming metropolitan scenes. The Brilliant Entryway Scaffold in San Francisco is one more striking illustration of iron-based structures that address designing and engineering ability. Steel-supported concrete, utilized in most present day structures, utilizes iron bars to give the strength expected to endure compressive and ductile powers.

The transportation area is another space vigorously dependent on iron-based materials. Autos, trains, boats, and planes generally integrate iron and steel to changing degrees. In the auto business, the utilization of iron and steel gives the fundamental strength and wellbeing highlights as case, body boards, and security confines. Rail transportation depends on iron-based tracks, wheels, and different parts, making it a necessary method of transportation universally. The sea business relies upon iron and steel for transport development, from freight vessels to military warships. Indeed, even in the flying area, where lightweight materials are valued, iron-based materials track down applications in motors, landing gear, and primary parts.

The development and support of streets, a key part of transportation foundation, include broad utilization of iron-based materials. Black-top, the essential material for street surfaces, is frequently blended in with totals containing iron metal. Iron and steel extensions and bridges are fundamental for exploring testing territory and interfacing networks. Guardrails, signs, and streetlamps, all produced using iron and steel, add to street wellbeing and proficient traffic the board.

Railroads, known as the "iron pony" in the nineteenth hundred years, changed travel, exchange, and the development of products. Iron rails, later built up by steel in current times, give the establishment to rail transport. Trains, carts, and other moving stock are transcendently made of iron and steel parts. The development of high velocity trains, which integrate progressed materials, actually depends on iron-rich tracks for solidness and wellbeing.

The sea business, imperative for worldwide exchange and transportation, essentially develops vessels from iron and steel. The toughness and consumption opposition of these materials pursue them ideal decisions for ships that should endure the brutal states of the vast ocean. Freight ships, luxury ships, and military vessels are totally developed with iron-based materials. The utilization of iron mineral, purified into steel, additionally reaches out to the formation of steel trailers, vital to the worldwide store network, empowering the proficient vehicle of products via ocean.

Aviation, an industry known for its dependence on lightweight materials, consolidates iron in fundamental parts. Stream motors, basic to the drive of current airplane, incorporate iron-based combinations that can endure high temperatures and tensions. Airplane landing gear, which should bear the weight and effect of departures and arrivals, is essentially made of iron and steel parts. The toughness and strength of these materials guarantee the wellbeing of air travel.

Iron likewise plays a fundamental job in the field of energy creation, from the development of force plants to the age of power. Iron and steel are urgent in the development of force plant foundation, like boilers, turbines, and generators. These materials can endure the outrageous circumstances and tensions related with the transformation of intensity into power. Electrical transformers, used to change voltage levels in the power matrix, contain iron centers that empower effective energy transmission.

The age of power from different sources depends on iron-based advancements. Coal-terminated power plants utilize iron boilers and turbines to change over the energy of consuming coal into power. Thermal energy stations additionally utilize iron and steel in their development, with the additional advantage of radiation protecting. In environmentally friendly power areas, iron and steel are basic to wind turbines, sun powered charger outlines, and hydroelectric generators. These advances outfit nature's assets to create clean energy while profiting from the sturdiness and erosion opposition of iron-based materials.

Iron's effect on regular daily existence reaches out to the domain of broadcast communications and data innovation. The development of electronic gadgets, including cell phones, PCs, and servers, includes the utilization of iron and steel in different parts. The housings of numerous electronic gadgets are developed from iron or steel for toughness and electromagnetic safeguarding. Iron-based materials are likewise utilized in the assembling of semiconductors, which are the structure blocks of current gadgets.

The worldwide media communications organization, which works with the trading of data across immense distances, depends on iron as fiber optic links. These links, encased in iron or steel sheaths, safeguard the fragile glass filaments that send information as beats of light. Iron's part in the media communications industry guarantees the fast transmission of data, associating people and organizations all over the planet.

Iron and steel keep on supporting basic framework, including water supply and sterilization frameworks. Water pipelines, answerable for the conveyance of clean water to homes and organizations, are transcendently developed utilizing iron and steel materials. These materials offer the fundamental strength and consumption protection from endure the afflictions of moving water. Likewise, sewage and wastewater treatment offices consolidate iron and steel parts to deal with the protected removal of waste and safeguard general wellbeing.

In the field of medical services, iron assumes an imperative part in the working of the human body. Iron is a fundamental dietary mineral that the body needs for various physiological cycles, remembering the vehicle of oxygen by hemoglobin for red platelets. A lack in dietary iron can prompt paleness, a condition portrayed by exhaustion, shortcoming, and decreased mental capability. Iron enhancements and iron-rich food varieties, like red meat, spinach, and strengthened oats, are urgent for keeping up with appropriate iron levels in the body.

Iron likewise assumes a significant part in the drug business, where it is utilized in the production of iron-based meds and enhancements. Iron enhancements are recommended to people with iron-inadequacy pallor or other ailments that require iron supplementation. Iron-based prescriptions are likewise utilized in the therapy of iron over-burden problems and certain ongoing illnesses.

The job of iron reaches out to the universe of style, where it is utilized in the coloring and shading of materials. Iron oxide shades, frequently got from iron mineral, are utilized to make a great many varieties in dress and texture coloring. These colors are utilized in both normal and manufactured colors, giving a range of tones in the design business.

The creation of family products and apparatuses intensely depends on iron and steel. Kitchenware, including pots, skillet, and utensils, is frequently developed from iron or steel for its toughness and intensity directing properties. Pressing sheets and irons, utilized for piece of clothing care, are fundamental things in numerous families. Pressing sheets, with iron racks made of iron or steel, give a steady surface to dress support.

Security and wellbeing are improved by the utilization of iron-based materials in lock and key frameworks. Iron keys and locks have a long history of giving security to homes, organizations, and important belongings. The utilization of iron in these frameworks is a demonstration of the solidness and dependability

6.3 Healthcare and Iron: From Supplements to MRI Machines

Medical care and medication have forever been characteristically connected to the regular components, and one of the most indispensable of these components is iron. Iron assumes a diverse part in medical services, reaching out from its presence as a fundamental micronutrient in the human body to its application in cutting edge clinical advances. This article digs into the assorted manners by which iron adds to the field of medical care, incorporating its effect on human science, clinical therapies, diagnostics, and the advancement of modern clinical gear.

The significance of iron in human wellbeing starts with its job as a basic supplement. Iron is a fundamental mineral that the body expects for different physiological capabilities, with its most notable capability being oxygen transport in the circulation system. Hemoglobin, a protein tracked down in red platelets, is the essential transporter of oxygen, and it contains iron at its center. At the point when we consume dietary iron, our body uses it to orchestrate hemoglobin, guaranteeing that oxygen is actually conveyed from the lungs to the body's tissues.

Iron's part in oxygen transport is urgent to generally prosperity. At the point when the body needs adequate iron, it can prompt iron-inadequacy sickliness, a condition portrayed by a diminished capacity to ship oxygen. Sickliness frequently appears with side effects like exhaustion, shortcoming, fair skin, and mental disabilities. Iron-lack paleness can happen because of deficient dietary admission, unfortunate iron ingestion, or conditions that lead to press misfortune.

Dietary iron exists in two structures: heme iron, which is found in creature items like meat and fish, and non-heme iron, which is available in both creature and plant-based food varieties. Heme iron is all the more promptly consumed by the body, while non-heme iron needs support from dietary variables like L-ascorbic acid for ideal assimilation. People with low dietary iron admission, veggie lovers, and those with conditions that influence iron ingestion are at higher gamble of creating iron-lack frailty. In such cases, iron enhancements, commonly as ferrous sulfate or ferrous gluconate, might be prescribed to recharge iron stores and right the sickliness.

Iron supplementation is a key part of medical services, especially for people who can't meet their iron requirements through diet alone. These enhancements are accessible in different structures, including tablets, containers, and fluid arrangements. They expect to furnish the body with the essential iron to deliver hemoglobin, improve oxygen transport, and reduce sickliness related side effects. Be that as it may, it is fundamental for medical services suppliers to painstakingly survey patients for iron-lack frailty and recommend supplements just when essential, as over the top iron admission can antagonistically affect wellbeing.

Past its part in the circulatory system, iron likewise assumes an essential part in the safe framework. The body's guard against contaminations and sicknesses depends on iron in white platelets, which are essential for the safe reaction. Iron assists white platelets with working really, permitting them to immerse and annihilate microbes, like microscopic organisms and infections.

Be that as it may, this equivalent dependence on iron is perceived by microbes, which likewise need iron to make due. Subsequently, the body's guideline of iron is a refined cycle, guaranteeing that iron is accessible for fundamental capabilities while limiting its accessibility to attacking microorganisms.

Medical services suppliers figure out the significance of iron guideline with regards to diseases. In specific sicknesses, similar to tuberculosis, mycobacteria flourish with iron inside the host's body. Consequently, overseeing iron levels in such contaminations turns into a procedure to restrict the endurance of microorganisms. Medical services experts might recommend iron chelation treatment, which includes controlling medications that tight spot to abundance iron in the body, making it less accessible to microbes. This technique assists the safe framework with combatting diseases all the more really.

Also, iron's commitment to human wellbeing reaches out to the neurological framework. Iron is critical for the legitimate turn of events and working of the mind. Deficient iron levels in the cerebrum can prompt mental debilitations and formative issues, especially in babies and small kids. Guaranteeing adequate iron admission, particularly during pregnancy and youth, is fundamental for the development of myelin, a substance that protects nerve strands, and for synapse creation.

Weakness because of lack of iron can likewise have neurological outcomes, prompting side effects, for example, trouble concentrating, memory issues, and mind-set problems. Research has shown that amending iron-lack paleness can prompt enhancements in mental capability and generally speaking prosperity. Therefore, medical services suppliers give close consideration to press status, particularly in weak populaces like pregnant ladies, newborn children, and youths.

The connection among iron and medical services further stretches out to the field of medication, where iron-based therapies are utilized to address different ailments. Iron imbuements, controlled intravenously, are a typical methodology for treating serious iron-lack frailty when oral enhancements are inadequate or ineffectively endured. These imbuements convey iron straightforwardly into the circulation system, bypassing the stomach related framework, and can quickly renew iron stores, prompting the adjustment of frailty.

Iron-chelating drugs, as referenced prior, have applications in the treatment of iron over-burden conditions. Hemochromatosis, an innate condition portrayed by extreme iron retention, frequently requires iron-chelation treatment to eliminate overabundance iron from the body and forestall organ harm. These prescriptions work by restricting to abundance iron and working with its discharge through pee or defecation. Standard checking of iron levels and close coordinated effort with medical care suppliers are fundamental for people with iron over-burden conditions.

Iron-based treatments additionally have applications in the therapy of specific persistent sicknesses. Constant kidney infection, for instance, can prompt pallor because of diminished erythropoietin creation, a chemical that invigorates red platelet creation. In such cases, medical services suppliers might endorse erythropoiesis-animating specialists (ESAs), which animate the bone marrow to create more red platelets. Some ESAs depend on recombinant human erythropoietin, while others are formed with iron, improving their viability in the therapy of pallor related with persistent kidney illness.

The impact of iron in medical services reaches out to the improvement of imaginative clinical advancements and demonstrative apparatuses. One of the most striking models is the attractive reverberation imaging (X-ray) machine, a central clinical imaging gadget utilized for diagnosing and observing different ailments. X-ray machines work on the standards of atomic attractive reverberation, which depends on the communication between hydrogen cores major areas of strength for and fields.

The attractive fields created by X-ray machines are delivered by superconducting magnets, which are cooled to incredibly low temperatures utilizing fluid helium.

The center of these superconducting magnets comprises of niobium-titanium (NbTi) amalgams, which are equipped for keeping up with superconducting properties in high attractive fields. Notwithstanding, to guarantee that these magnets are charged and kept up with at the vital low temperatures, they require iron-center magnets, called resistive magnets. Iron is a superb material for this reason in view of its attractive properties and high warm conductivity. These resistive magnets produce attractive fields for charging the superconducting magnets, making X-ray filters conceivable.

X-ray machines have reformed medical care by giving itemized pictures of the body's inward designs without utilizing ionizing radiation. They are generally utilized in the finding of conditions going from mind growths and joint wounds to coronary illness and disease. Their painless nature and capacity to catch high-goal pictures have made X-ray examines a basic instrument for medical services suppliers, empowering precise judgments and therapy arranging.

The utilization of iron isn't restricted to the gear that upholds clinical imaging. Iron oxide nanoparticles, small particles of iron, have acquired noticeable quality in the field of diagnostics and treatment. These nanoparticles have attractive properties that make them reasonable for attractive reverberation imaging. In any case, their applications reach out past imaging.

Chapter 7

The Legacy Continues

As the world advances into the 21st 100 years, the tradition of iron is a long way from blurring. Iron, perhaps of the most plentiful and adaptable component on The planet, keeps on forming human progress, mechanical headways, and daily existence in horde ways. Its commitments range across ventures, from development and transportation to medical care and advancement. This paper investigates the persevering through tradition of iron, stressing its continuous importance in our advancing world.

The Job of Iron in Design and Development

Iron has been a vital piece of building and development attempts for a really long time, and it stays fundamental in molding our fabricated climate. The improvement of elevated structures, many-sided spans, and tough foundation relies upon iron and its composite, steel. The tradition of iron in design is exemplified by its job in famous designs like the Burj Khalifa in Dubai, remaining as the tallest structure on the planet and developed with a steel outline. Additionally, the Millau Viaduct in France, praised for its designing brightness, consolidates immense amounts of iron and steel, highlighting the material's importance in current development.

Present day development strategies and techniques have outfit the strength and solidness of iron and steel to make dazzling designs as well as protected and strong ones. Supported concrete, a blend of steel and cement, is at the center of contemporary development. Steel support bars, known as rebars, give elasticity, upgrading substantial's capacity to endure loads and oppose breaking. This innovation is omnipresent in the development of high rises, parkways, and other foundation. Iron's commitment to development materials guarantees that these designs stay steady and utilitarian, making urbanization conceivable on a fabulous scale.

Iron in Transportation: A Consistently Developing Excursion

The transportation business has been irreversibly formed by iron, and it keeps on developing with innovative headways. The tradition of iron in transportation envelops a wide range, from the primary steam trains and steamships to present day cars, trains, airplane, and rocket.

The coming of steam trains in the nineteenth century denoted a transformation in transportation, frequently alluded to as the "Iron Pony" period. Iron and steel were the chief materials utilized in the development of trains and rail line tracks. As the rail route networks extended across landmasses, the transportation of products and individuals turned out to be more effective and open, moving the Modern Transformation.

The iron and steel industry gave the spine to rail route development, with rail tracks and trains representing the time of modern advancement. Indeed, even today, railroads stay a vital part of worldwide transportation frameworks, underlining iron's continuous importance. Fast trains, known for their effectiveness and maintainability, keep on depending on iron tracks and wheels, interfacing urban areas and countries with wonderful speed and accuracy.

The oceanic business, one more domain where iron's inheritance wins, keeps on depending on iron and steel for transport development. From freight vessels and traveler liners to military warships, the utilization of these materials guarantees sturdiness and life span adrift. Compartment ships, major to worldwide exchange, likewise rely upon iron-based materials, further exhibiting iron's job in forming current economies.

Flying and aviation, famous for their accentuation on accuracy and security, keep on integrating iron-based materials into the development of airplane and shuttle. Fly motors, fundamental to the impetus of present day airplane, utilize iron-based composites known for their capacity to endure high temperatures and tensions. Airplane landing gear, which bears the weight and effect of departures and arrivals, comprises fundamentally of iron and steel parts, underlining their basic job in guaranteeing the security of air travel. Furthermore, rockets, which send off satellites into space and convey space explorers to the Global Space Station, rely upon the dependability of iron-based materials.

Iron's effect ashore, ocean, and sky is increased by the thriving field of electric and independent vehicles. While the materials utilized in these vehicles shift, iron remaining parts applicable in the creation of electric engines and parts. Zap of transportation, in a bid to lessen ozone harming substance discharges and dependence on petroleum derivatives, shows the proceeded with job of iron in molding the fate of versatility.

Iron in Energy Age and Capacity

Iron's commitment to the age and capacity of energy is significant in tending to contemporary energy needs and ecological worries. From power plants to batteries, iron-based advancements drive progress in energy creation, stockpiling, and manageability.

The development of force plants depends on iron and steel, especially in basic parts like boilers, turbines, and generators. These materials can endure outrageous circumstances and tensions, which are innate in the change of intensity into power. Iron is fundamental in this unique circumstance, guaranteeing the proficiency and dependability of energy age from sources like coal, petroleum gas, and atomic parting.

In environmentally friendly power areas, iron and steel assume a critical part. Wind turbines, fundamental for bridling wind energy, require iron parts in their development. Iron is available in the casings, pinnacles, and gearboxes of wind turbines, guaranteeing their steadiness and execution. Also, sun powered charger outlines integrate iron-based materials that help photovoltaic modules, empowering the saddling of sun oriented energy. Hydroelectric power age, another environmentally friendly power source, benefits from iron and steel parts in turbines and generators, working with the transformation of water stream into power.

Power transmission and dissemination include the utilization of iron centers in transformers, gadgets that change voltage levels in the power lattice. These iron centers upgrade the proficiency of energy transmission, guaranteeing that power can be conveyed across significant distances with insignificant misfortunes. Iron's attractive properties are essential to the activity of transformers, highlighting its persevering through job in the energy area.

The improvement of energy stockpiling advancements has acquired noticeable quality as society tries to adjust fluctuating sustainable power sources with predictable energy supply. Iron-based batteries, for example, the lithium iron phosphate (LiFePO4) battery, offer a promising arrangement. These batteries are prestigious for their security, long cycle life, and ecological amicability. Iron's plentiful accessibility and minimal expense add to the possibility of far reaching energy stockpiling, making it a necessary piece of the progress to sustainable power frameworks.

Iron's Effect on Medical care: From Enhancements to X-ray Machines

Iron's pertinence in medical services is multi-layered and extensive, enveloping the two its fundamental job in human science and its applications in clinical therapies and high level clinical advances.

As a fundamental mineral, iron holds a vital spot in human wellbeing. The human body requires iron for different physiological capabilities, with its most perceived job being oxygen transport in the circulatory system. Hemoglobin, the iron-containing protein tracked down in red platelets, is liable for conveying oxygen from the lungs to the body's tissues.

Iron-inadequacy weakness, a condition portrayed by a diminished ability to move oxygen, can happen because of lacking dietary admission, unfortunate iron ingestion, or conditions that lead to press misfortune.

The results of pallor are exhaustion, shortcoming, fair skin, and mental debilitations. To address this, medical care suppliers recommend iron enhancements as ferrous sulfate or ferrous gluconate. These enhancements renew iron stores and right the paleness, working on the patient's general wellbeing and prosperity.

Past oxygen transport, iron assumes a vital part in the safe framework. Iron is imperative for the working of white platelets, which are vital to the body's safeguard against contaminations and sicknesses. Iron assists white platelets with working productively, permitting them to inundate and obliterate microorganisms like microbes and infections. Iron's presence is likewise perceived by microbes, which seek iron in the body to work with their own endurance. In this manner, the body's guideline of iron is a mind boggling process that guarantees its accessibility for fundamental capabilities while restricting its openness to attacking microorganisms.

In medical care, overseeing iron levels can be an essential way to deal with tending to contaminations. In specific sicknesses like tuberculosis, mycobacteria depend on iron inside the host's body for endurance. Subsequently, medical services experts might endorse iron chelation treatment, which includes the organization of medications that tight spot to overabundance iron in the body, restricting its accessibility to microbes. This procedure supports the resistant framework's capacity to successfully battle diseases.

Iron's impact on medical services reaches out to the therapy of explicit ailments. Iron mixtures, controlled intravenously, are a typical methodology for treating serious iron-inadequacy sickliness when oral enhancements demonstrate lacking. Iron implantations furnish the body with the important iron to create hemoglobin, improving oxygen transport and lightening weakness related side effects. These intravenous imbuements are critical for people who can't meet their iron necessities through diet or oral supplementation alone.

The utilization of iron-chelating drugs likewise has applications in the treatment of iron over-burden conditions. Inherited hemochromatosis, portrayed by exorbitant iron assimilation, frequently requires iron-chelation treatment to eliminate abundance iron from the body and forestall organ harm. These drugs work by restricting to overabundance iron and working with its excretion.

7.1 Iron's Enduring Influence

Iron, the bountiful and flexible component, has made a permanent imprint on human development, impacting bunch parts of our lives. Its getting through impact ranges across history, industry, innovation, and even wellbeing, helping us to remember the key job iron plays in our cutting edge world. This article dives into the diverse manners by which iron keeps on shaping our lives, underlining its consistently present importance and advancement.

Verifiable Establishments: Iron as the Foundation of Development

The verifiable underpinnings of iron's impact are well established in human progress. Iron has been instrumental in molding the course of history, especially through its utilization in weaponry, devices, and foundation.

Iron's use in weaponry reformed the idea of fighting. The Iron Age, set apart by the broad utilization of iron devices and weapons, supplanted the prior Bronze Age. Iron's hardness and sturdiness brought about more effective and deadly weaponry, prompting massive changes in military system and strategies. The tradition of iron in weaponry can be followed through the ages, from iron swords and lances to guns and guns.

The Roman Domain, prestigious for its designing and development accomplishments, utilized iron in its foundation. Iron nails, for example, assumed an essential part in the development of the notorious Roman streets, extensions, and water channels. The Roman accentuation on strong iron foundation added to the life span of their engineering wonders, some of which keep on remaining right up 'til now.

Iron's Part in Industry and Modernization

The modern unrest of the eighteenth and nineteenth hundreds of years denoted a defining moment in mankind's set of experiences, and iron was at the very front of this extraordinary time. The tradition of iron in industry and modernization is significant, as it empowered the motorization and large scale manufacturing that portrayed this period.

The improvement of steam motors, fueled by iron boilers and apparatus, altered industry and transportation. Iron's part in steam power can be exemplified by the steam train, which prodded the development of railways and worked with the development of products and individuals on an exceptional scale. The development of ironclad boats, safeguarded by iron shield, denoted a critical headway in maritime innovation, with these vessels overwhelming maritime fighting.

Iron likewise assumed a critical part in the rise of plants and assembling processes. Iron and steel were utilized to build the machines and devices essential for modern creation. The tradition of iron in industry is clear in the development of enormous scope processing plants, stockrooms, and modern edifices that fueled the creation of materials, hardware, and different products.

The tradition of iron in industry proceeds to the current day, as iron and steel stay the essential materials for assembling and development. Steel, a powerful iron combination, is fundamental for developing current structures, scaffolds, and framework. It is the foundation of high rises, adding to their solidarity and steadiness. In the auto business, iron and steel are utilized widely for vehicle edges, motors, and wellbeing highlights. As a matter of fact, the car business is one of the biggest customers of steel universally.

Iron's job in transportation stretches out to vehicles, trains, boats, and airplane. Iron and steel are utilized in different parts, from vehicle casings to rail tracks, transport frames, and airplane parts. These materials add to the strength, solidness, and security of transportation frameworks around the world.

Iron in Framework and Urbanization

Iron's commitment to urbanization is huge and expansive, molding current urban communities and networks. Metropolitan framework, which envelops structures, streets, extensions, and utilities, depends on iron and steel to give strength and life span.

Steel-supported concrete, a basic development material, integrates iron poles to improve its elasticity. This advancement has been critical in developing elevated structures and extensive framework. Notorious designs like the Domain State Working in New York City and the Brilliant Door Extension in San Francisco stand as demonstrations of iron's persevering through job in forming metropolitan scenes.

Transportation framework, another space intensely dependent on iron-based materials, incorporates rail lines, streets, and extensions. Iron and steel extensions and bridges are fundamental for exploring testing territory and interfacing networks. Rail lines, known as the "iron pony" in the nineteenth 100 years, upset travel, exchange, and the development of products. Rail tracks and moving stock are dominatingly made of iron and steel parts.

In the oceanic business, iron and steel are the essential materials for transport development, from freight vessels to military warships. Iron's solidness and erosion opposition pursue it the ideal decision for ships that should endure the brutal states of the vast ocean. Iron-based materials are additionally utilized in the production of steel trailers, vital to the worldwide store network, empowering the proficient vehicle of merchandise via ocean.

Aviation, an industry known for its dependence on lightweight materials, consolidates iron in fundamental parts. Fly motors, basic to the drive of present day airplane, incorporate iron-based amalgams that can endure high temperatures and tensions. Airplane landing gear, which should bear the weight and effect of departures and arrivals, is fundamentally made of iron and steel parts. The toughness and strength of these materials guarantee the wellbeing of air travel.

In the field of medical care, iron assumes a crucial part in the working of the human body. Iron is a fundamental dietary mineral that the body needs for various physiological cycles, remembering the vehicle of oxygen by hemoglobin for red platelets. A lack in dietary iron can prompt frailty, a condition described by exhaustion, shortcoming, and diminished mental capability. Iron enhancements and iron-rich food sources, like red meat, spinach, and strengthened oats, are significant for keeping up with appropriate iron levels in the body.

Iron likewise assumes a significant part in the drug business, where it is utilized in the production of iron-based meds and enhancements. Iron enhancements are endorsed to people with iron-lack sickliness or other ailments that require iron supplementation. Iron-based prescriptions are likewise utilized in the therapy of iron over-burden problems and certain constant illnesses.

The job of iron reaches out to the universe of style, where it is utilized in the coloring and shading of materials. Iron oxide shades, frequently got from iron metal, are utilized to make a large number of varieties in dress and texture coloring. These shades are utilized in both regular and manufactured colors, giving a range of tones in the design business.

The creation of family merchandise and machines intensely depends on iron and steel. Kitchenware, including pots, skillet, and utensils, is frequently built from iron or steel for its solidness and intensity leading properties. Pressing sheets and irons, utilized for piece of clothing care, are fundamental things in numerous families. Pressing sheets, with iron racks made of iron or steel, give a steady surface to dress upkeep.

Security and wellbeing are upgraded by the utilization of iron-based materials in lock and key frameworks. Iron keys and locks have a long history of giving security to homes, organizations, and significant belongings. The utilization of iron in these frameworks is a demonstration of the toughness and dependability of the material.

The development and upkeep of streets, a major part of transportation framework, include broad utilization of iron-based materials. Black-top, the essential material for street surfaces, is frequently blended in with totals containing iron metal. Iron and steel scaffolds and bridges are fundamental for exploring testing landscape and interfacing networks. Guardrails, signs, and streetlamps, all produced using iron and steel, add to street security and effective traffic the board.

Iron in Medical services: From Enhancements to X-ray Machines

The impact of iron in medical services reaches out to the domain of media communications and data innovation. The creation of electronic gadgets, including cell phones, PCs, and servers, includes the utilization of iron and steel in different parts. Iron and steel housings give strength and security to delicate electronic hardware. The tradition of iron in data innovation is obvious in the dependability and vigor of current gadgets, adding to their far and wide use.

The field of clinical innovation benefits from iron's attractive properties and conductivity. Iron oxide nanoparticles, minuscule iron-based particles, have acquired noticeable quality in diagnostics and treatment. In attractive reverberation imaging (X-ray), iron oxide nanoparticles are utilized as differentiation specialists, upgrading the perceivability of explicit tissues and designs in the body. This application has fundamentally worked on the indicative capacities of X-ray, giving itemized pictures of the mind, heart, and different organs. Iron's attractive reverberation imaging (X-ray), which depends major areas of strength for on fields produced by superconducting magnets, has turned into a foundation of present day clinical diagnostics. X-ray machines produce high-goal pictures of the body's inward designs, supporting the determination of different ailments. The center of superconducting magnets, utilized in X-ray machines, comprises of niobium-titanium (NbTi) composites, equipped for keeping up with superconducting properties in high attractive fields.

To charge and keep up with these superconducting magnets at low temperatures utilizing fluid helium, iron-center magnets, called resistive magnets, are utilized.

Iron's attractive properties and high warm conductivity make it an ideal material for these resistive magnets, which are essential to X-ray innovation.

7.2 The Ongoing Innovations and Future Prospects

As we explore the 21st 100 years, iron's persevering through inheritance proceeds to advance and adjust to the steadily changing scene of human development. This article investigates the continuous advancements and future possibilities of iron's impact in different fields, from industry and innovation to medical services and natural maintainability.

Iron in Industry: Developments and Manageability

In the domain of industry, iron remaining parts a foundation of assembling and development, with continuous developments pointed toward improving its maintainability and effectiveness. The creation of steel, a basic iron compound, has seen critical progressions as of late, zeroing in on diminishing natural effect and upgrading asset use.

One such advancement is the improvement of electric circular segment heaters (EAFs) for steel creation. EAFs use power to liquefy scrap steel, offering an additional energy-proficient and harmless to the ecosystem option in contrast to conventional impact heaters. This innovation lessens ozone depleting substance discharges and limits the utilization of regular assets. EAFs are especially imperative for reusing steel, as they can effectively liquefy and refine scrap steel, expanding the existence pattern of this fundamental material.

Developments in metallurgy have prompted the production of cutting edge high-strength prepares, customized for explicit applications. These prepares join strength and solidness with diminished weight, making them ideal for the car business, where eco-friendliness and wellbeing are central. High-strength prepares empower the plan of vehicles that are both lightweight and powerful, adding to diminished fuel utilization and improved traveler security.

The utilization of iron and steel in development has seen a resurgence in imaginative methodologies, for example, the idea of "green structures." These harmless to the ecosystem structures consolidate iron and steel in their edges and parts while utilizing reasonable structure rehearses. Steel-outlined structures are known for their sturdiness and recyclability. Also, the adaptability of iron-based materials has empowered the advancement of tremor safe structures, guaranteeing the security of metropolitan populaces in seismic locales.

Iron's job in the development business reaches out to savvy and supportable structures. Current development techniques, like measured and pre-assembled development, depend on steel casings and parts. These creative procedures smooth out development processes, decrease waste, and upgrade energy proficiency.

Moreover, canny structure materials, incorporating those with iron-based properties, can possibly change how structures are planned and worked. These materials can effectively answer ecological circumstances, self-fix, and further develop energy productivity.

The utilization of iron and steel in the development of extensions and framework additionally keeps on advancing. Creative materials and designing procedures have prompted the improvement of elite execution and enduring extensions. Consumption safe coatings and high level underlying models have expanded the life expectancy of iron-based structures, decreasing upkeep costs and improving security.

Iron in Transportation: Manageable Versatility

In the transportation area, the continuous developments including iron and steel add to more feasible and proficient portability arrangements. The drive for diminished emanations and expanded energy proficiency has prompted headways in vehicle plan, materials, and impetus advancements.

Electric vehicles (EVs) address a huge change in the auto business, with iron and steel assuming fundamental parts in their creation. EVs utilize iron-based materials in electric engines, casings, and batteries. Iron and steel parts offer strength and sturdiness, making them reasonable for vehicle development. Developments in battery innovation, including lithium iron phosphate (LiFePO4) batteries, have improved the energy stockpiling limit and security of EVs, further advancing their reception as manageable transportation arrangements.

Iron and steel are additionally integral to the improvement of rapid rail frameworks, like maglev (attractive levitation) trains. These high level transportation frameworks depend major areas of strength for on based materials for their tracks, which work with high velocity travel with insignificant erosion and commotion. Maglev innovation further develops transportation productivity as well as lessens the ecological effect of rail travel.

The aeronautic trade keeps on enhancing with iron-based materials in airplane development. Headways in aeronautics innovation have prompted the advancement of additional eco-friendly and harmless to the ecosystem airplane. The utilization of cutting edge combinations, including those containing iron, offers a harmony among strength and weight. Besides, plane design specialists are exploring iron-based materials that can endure outrageous circumstances, for example, those experienced in hypersonic flight, opening additional opportunities for rapid air travel.

Iron's impact in sea transportation stretches out to economical arrangements in shipbuilding. Developments incorporate the improvement of more eco-friendly boat plans and the utilization of cutting edge iron-based coatings to lessen grinding and lower fuel utilization.

Moreover, the investigation of elective powers, like alkali or hydrogen, is supposed to prompt the development of eco-accommodating vessels that utilization iron and steel in their development while diminishing ozone depleting substance emanations.

Iron and Maintainable Energy

The 21st century is portrayed by a developing obligation to economical energy arrangements, and iron is assuming an essential part in this progress. From environmentally friendly power age to energy capacity innovations, iron-based materials are indispensable to the improvement of a cleaner and more reasonable energy scene.

The development of power from sustainable sources depends on iron and steel parts in different ways. Wind turbines, fundamental for tackling wind energy, consolidate iron materials in their development. The casings, pinnacles, and gearboxes of wind turbines are made of iron and steel, guaranteeing their solidness and execution. Sun powered chargers, one more wellspring of environmentally friendly power, utilize iron-based materials in their edges and supports, improving the strength and effectiveness of these energy frameworks.

Hydroelectric power age, which tackles the energy of streaming water, relies upon iron and steel parts in its turbines and generators. These materials are fundamental for changing over water stream into power, adding to the creation of perfect and environmentally friendly power. Hydropower is a foundation of reasonable energy arrangements around the world, furnishing a dependable wellspring of power with insignificant ozone harming substance discharges.

Iron's part in energy stretches out to the transmission and appropriation of power. Transformers, fundamental gadgets in the power lattice, utilize iron centers to change voltage levels, guaranteeing effective power transmission over significant distances. These iron centers upgrade the exhibition of transformers, empowering the solid inventory of power to homes and organizations. Iron's attractive properties are significant to the activity of transformers, making it an irreplaceable material in the electrical business.

The advancement of energy stockpiling innovations has become progressively vital to adjust the irregular idea of sustainable power sources with steady energy supply. Iron-based batteries, for example, lithium iron phosphate (LiFePO4) batteries, have acquired conspicuousness as a feasible answer for energy capacity. These batteries are known for their wellbeing, long cycle life, and ecological neighborliness. Iron's bountiful accessibility and minimal expense add to the possibility of far and wide energy stockpiling, making it a necessary piece of the change to sustainable power frameworks.

One more road of development in energy capacity includes iron-based stream batteries. These frameworks utilize iron particles to store energy in fluid electrolytes.

Stream batteries offer benefits concerning versatility, long cycle life, and the capacity to store energy for broadened periods. They are especially reasonable for network scale energy capacity and the reconciliation of environmentally friendly power sources into the power framework.

As the world wrestles with the difficulties of environmental change and the requirement for manageable energy arrangements, iron will keep on assuming a significant part in driving the progress to cleaner and more proficient energy frameworks. Advancements in materials and innovations will additionally improve iron's commitments to sustainable power age and energy stockpiling.

Iron in Medical services: Propelling Diagnostics and Therapies

The medical services area is ready for critical headways, driven by developments in iron-based advances and materials. The continuous advancement in medical services and clinical sciences holds the commitment of further developed diagnostics, therapies, and patient consideration.

In the field of diagnostics, iron oxide nanoparticles are arising as amazing assets for imaging and designated drug conveyance. These nanoparticles can be directed to explicit tissues or organs in the body, offering exact symptomatic abilities and the potential for limited medicines. Attractive reverberation imaging (X-ray), which uses iron oxide nanoparticles as difference specialists, is supposed to turn out to be significantly more precise and educational, furnishing medical services suppliers with definite experiences into a patient's condition.

Additionally, iron-based advancements are opening additional opportunities for the discovery of illnesses and conditions at a prior stage. Attractive biosensors, which utilize iron nanoparticles, can identify infection markers with high awareness and particularity. These innovations offer harmless and fast indicative arrangements, empowering prior intercessions and working on tolerant results.

7.3 Reflection on Iron's Legacy in Innovation

The tradition of iron in development is a demonstration of the significant effect of an apparently unassuming component on the course of human progress. Since forever ago, iron has made a permanent imprint on different aspects of society, from industry and innovation to medical services and maintainability. Pondering iron's heritage in advancement, one is struck by the getting through impact of this flexible material and its capacity to adjust and develop because of the always changing necessities of humankind.

Verifiable Importance: Iron as an Impetus for Change

The verifiable meaning of iron as an impetus for change is promptly evident. The change from the Bronze Age to the Iron Age denoted a critical crossroads in mankind's set of experiences. Iron's hardness and sturdiness achieved an upheaval in weaponry, prompting huge changes in military procedure and fighting.

The tradition of iron in this setting is obvious in the change of swords and lances into additional deadly and productive weapons, adding to the ascent and fall of realms.

Also, iron's part in framework and development was vital in the improvement of old human advancements. The Roman Realm, eminent for its building wonders, utilized iron nails in the development of streets, scaffolds, and water systems. These getting through structures are a demonstration of the strength and life span of iron-based materials.

Industry and Modernization: Iron as the Foundation of Progress

The modern transformation of the eighteenth and nineteenth hundreds of years denoted a huge change in the utilization of iron as the foundation of progress. The motorization and large scale manufacturing made conceivable by iron and steel introduced another period of industrialization. The approach of steam motors, controlled by iron boilers and hardware, changed transportation and assembling.

Ironclad boats, safeguarded by iron shield, reformed maritime fighting and denoted a critical progression in military innovation. Iron and steel turned into the fundamental materials for the development of huge scope processing plants, stockrooms, and modern edifices, further driving the modern insurgency.

The tradition of iron in industry proceeds right up to the present day. Steel, a powerful iron combination, stays an essential material for assembling and development. It is indispensable to the development of present day structures, scaffolds, and framework, filling in as the foundation of high rises and other compositional wonders. The car business, one of the biggest customers of steel, depends on iron and steel broadly for vehicle casings, motors, and wellbeing highlights.

In the transportation area, iron-based materials are utilized in different parts of vehicles, trains, boats, and airplane. These materials add to the strength, solidness, and wellbeing of transportation frameworks around the world. The utilization of iron and steel in street development and transportation framework, like railroads and scaffolds, guarantees protected and effective travel.

Iron in Framework and Urbanization: Molding Current Urban areas

Iron's job in framework and urbanization significantly affects molding current urban communities and networks. Metropolitan framework, including structures, streets, extensions, and utilities, depends on iron and steel to give strength and life span.

Steel-built up concrete, a basic development material, integrates iron poles to improve its rigidity. This development has been urgent in building elevated structures, broad extensions, and other basic foundation. Famous designs like the Realm State Working in New York City and the Brilliant Entryway Scaffold in San Francisco stand as models of iron's persevering through job in molding metropolitan scenes.

Transportation framework, including railroads, streets, and scaffolds, depends on iron and steel materials to interface networks and explore testing territory. The development of iron-based scaffolds and bridges, joined with cutting edge designing, adds to productive and safe transportation organizations. Guardrails, signs, and streetlamps, produced using iron and steel, improve street security and traffic the executives.

In the oceanic business, iron and steel are the essential materials for transport development, with their strength and consumption obstruction pursuing them ideal decisions for vessels that should endure the cruel states of the vast ocean. Steel trailers, essential to the worldwide inventory network, utilize iron-based materials for proficient freight transport.

Aviation, an industry known for its dependence on lightweight materials, consolidates iron in fundamental parts. Stream motors, landing gear, and underlying parts of airplane use iron-based combinations, with their solidness and strength guaranteeing the security of air travel. Aviation innovation, driven by developments in iron-based materials, has extended the potential outcomes of rapid air travel and investigation.

In medical care, iron assumes an imperative part in the working of the human body. It is a fundamental dietary mineral, urgent for different physiological cycles, remembering the vehicle of oxygen by hemoglobin for red platelets. The accessibility of iron enhancements and iron-rich food sources is fundamental for keeping up with appropriate iron levels in the body.

In the drug business, iron is utilized in the production of iron-based prescriptions and enhancements. Iron enhancements are recommended to people with iron-lack frailty or other ailments that require iron supplementation. Iron-based prescriptions are additionally utilized in the therapy of iron over-burden problems and certain ongoing sicknesses.

The creation of family merchandise and apparatuses intensely depends on iron and steel. Kitchenware, including pots, skillet, and utensils, is frequently developed from iron or steel for their strength and intensity leading properties. Pressing sheets and irons, utilized for piece of clothing care, are fundamental things in numerous families, giving strength and proficiency in apparel upkeep.

Security and wellbeing are improved by the utilization of iron-based materials in lock and key frameworks. Iron keys and locks have a long history of giving security to homes, organizations, and important belongings. The utilization of iron in these frameworks is a demonstration of the strength and unwavering quality of the material.

Advancement in Transportation: Iron's Part in Supportable Portability

In the 21st hundred years, the transportation area is going through an extraordinary shift toward supportable portability. Iron and steel keep on assuming basic parts in these advancements, adding to more energy-effective and eco-accommodating transportation arrangements.

Electric vehicles (EVs) are at the front of economical portability, with iron and steel materials utilized in their creation. EVs influence iron-based parts in electric engines, edges, and batteries, offering strength, sturdiness, and solidness. Advancements in battery innovation, including lithium iron phosphate (LiFePO4) batteries, have further developed energy capacity limit and wellbeing, advancing the reception of EVs as harmless to the ecosystem transportation choices.

Rapid rail frameworks, like maglev (attractive levitation) trains, rely areas of strength for upon based materials for their tracks, empowering frictionless high velocity travel. Maglev innovation upgrades transportation effectiveness and limits the ecological effect of rail travel.

The airplane business is advancing with iron-based materials in airplane development, adding to the improvement of more eco-friendly and eco-accommodating airplane. Progressed composites, including those containing iron, offer a harmony among strength and weight, prompting further developed eco-friendliness and decreased natural effect.

In oceanic transportation, supportable arrangements in shipbuilding are turning into a concentration, with developments, for example, more eco-friendly boat plans and high level iron-based coatings that lessen erosion and fuel utilization. The investigation of elective fills, for example, alkali or hydrogen, guarantees eco-accommodating vessels built with iron and steel parts.

Iron in Medical services: Advances in Diagnostics and Therapy

In medical services, iron-based advancements keep on propelling diagnostics and therapies, offering the commitment of worked on understanding consideration. Iron oxide nanoparticles have arisen as integral assets for imaging and designated drug conveyance. Directed to explicit tissues or organs, these nanoparticles empower exact diagnostics and limited medicines. Attractive reverberation imaging (X-ray), using iron oxide nanoparticles as differentiation specialists, is turning out to be more exact and educational, giving itemized bits of knowledge into patients' circumstances.

Advancements in iron-based treatments are changing medical care. Iron imbuement methods have advanced to upgrade the conveyance of iron to patients, further developing therapy results for those with iron-inadequacy pallor and other ailments. Novel iron chelators offer more viable and designated therapies for iron over-burden problems and certain persistent illnesses.

The eventual fate of medical services guarantees customized medication, where iron and different therapies are custom-made to people in light of their hereditary and sub-atomic profiles. Advancements in regenerative medication and tissue designing, utilizing iron-based materials, offer the potential for biocompatible and bioactive frameworks for tissue fix and recovery.

Difficulties and Amazing open doors for What's in store

The fate of iron-related enterprises presents the two difficulties and amazing open doors. Guaranteeing capable and economical iron metal extraction is basic. Feasible mining rehearses, decreased natural effects, and dependable asset the executives are fundamental for the moral and earth cognizant creation of iron.

Asset productivity is a basic concentration for what's in store. Developments in reusing and round economy practices can decrease the interest for essential iron metal extraction and cutoff the ecological impression of iron-based materials. The improvement of eco-accommodating.